The Big Smoke

Norman fireplace

The Big Smoke

A HISTORY OF
AIR POLLUTION IN LONDON
SINCE MEDIEVAL TIMES

Peter Brimblecombe

Methuen

LONDON AND NEW YORK

First published in 1987 by
Methuen & Co. Ltd
11 New Fetter Lane, London EC4P 4EE

Published in the USA by
Methuen & Co.
in association with Methuen, Inc.
29 West 35th Street
New York NY 10001

© *1987 Peter Brimblecombe*

Typeset in Monophoto Garamond by
Vision Typesetting, Manchester

Printed in Great Britain at the
University Press, Cambridge

British Library Cataloguing in Publication Data

Brimblecombe, Peter
The big smoke: a history of air pollution
in London since medieval times.
1. Air – Pollution – England –
London – History
I. Title
363.7'392'09412   TD883.7.G72L6

ISBN 0-416-90080-1

Library of Congress Cataloging in Publication Data

Brimblecombe, Peter.
The big smoke.
Bibliography: p.
Includes index.
1. Air – Pollution – England – London Metropolitan
Area – History.   I. Title.
TD883.7.G72L663 1987   363.7'392'09412   86-18003
ISBN 0-416-90080-1

Contents

List of figures and sources

List of tables

Acknowledgements

I would like to thank all those who were wide-eyed with amazement at the idea that there may have been air pollution in 'the good old days'. It is their interest that has kept me going all these years. I am particularly grateful to Professor Hubert Lamb for reading a disastrous early draft and Frances Nicholas for aiding its conversion to a more acceptable second draft. Later on I found comments from reviewers such as Lord Ashby and Mr James R. Sewell particularly useful. I thank the typists who grappled with the manuscript in the days before word processors: Jill Newham, Julie Fox, Katherine Rutherford, Joy Leeds, Jane Horsfall. And Bach and Beethoven, whose music soothed the hours.

The author and publishers are grateful to Faber and Faber Limited for granting permission to reprint 20 lines from 'Glyn Cynon Wood' from *Welsh Poems: Sixth Century to 1600* translated by Gwyn Williams.

I
History and early air pollution

History is junk![1] A society may be characterized by the contents of its drains, waste piles, water-closets, graveyards or chimneys. Smoke and soot have generally been considered too tenuous for archaeological examination, and the historian has usually held an orange before his nose and made but fleeting reference to fleas, fumes, sweat and dung. An irreverent few have handled the less polite aspects of human society. They have pondered the contents of rubbish dumps and viewed life from a medieval latrine, but the way in which our ancestors fouled the air has usually been neglected. Being irreverent, and not a historian, I have decided to write London's history from this latter perspective.

This is not as difficult as might be imagined because even the earliest societies had no difficulty in causing a wide variety of environmental damage. They polluted the air and waters, destroyed soils and eradicated flora and fauna. We seem to excuse this behaviour because the magnitude of the changes they wrought were usually small and many people currently see an innocence in their actions that makes them more permissible than those of urbanized societies. Besides, time has changed much of the environmental damage of the past into the archaeological and recreational sites of today. In Norfolk, early mining left us the Broads and Grimes Graves near Thetford. A society requires no special schooling to learn how to damage the environment, but concern for resources is often imposed in the form of religious principles, morals or taboos. It seems as if the desire to expend meagre resources to preserve environmental quality is not necessarily innate. It is true that many cultures have emphasized the need to live in harmony with the environment, but even where this is keenly felt, as in oriental societies, environmental ideals have often been lost in the face of material needs.[2]

Interior air pollution

It is not hard to visualize the way in which the earliest air pollution occurred. The smoke from fires within huts would have filled the whole

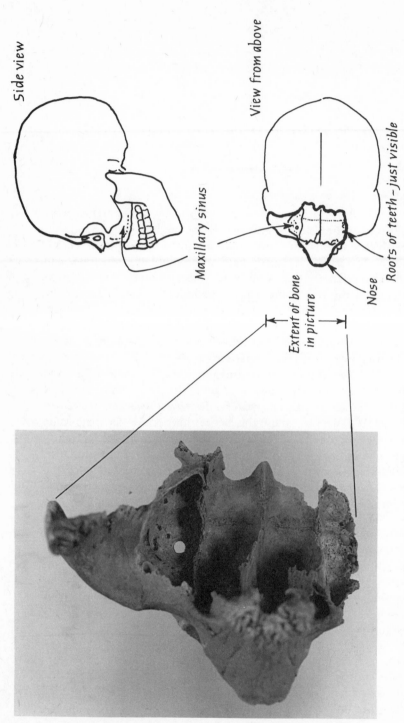

Side view

Maxillary sinus

View from above

Roots of teeth – just visible

Nose

Extent of bone
in picture

FIGURE 1.1 Fragment of a late Saxon/early medieval skull from Norwich. Note pitting and perforation to the floor of the sinus which probably arose from prolonged sinusitis (Skeletal material HOLE 25 kindly lent by Castle Museum, Norwich. Photographed by Stuart Robinson.) *and* Sketches of human skull showing location of maxillary sinus and area covered by skull fragment

interior before finding its way out through a hole in the roof. The removal of smoke was difficult to make effective because any hole big enough to remove all the smoke let in the rain as well. Some of the arrangements for ventilating early dwellings were ingeniously designed,[3] but there is a large amount of archaeological evidence which convinces one that indoor pollution must have been rather troublesome to early man.

The best evidence for interior air pollution in the distant past comes from examination of samples of lung tissue. These have been preserved by freezing or dessication. Freezing is usually accidental, but dessication or mummification is often deliberate. The fact that these two processes occur at opposite climatic extremes means that materials from quite diverse populations are available for palaeopathological examination. Blackening of lung tissues through long exposure to smoky interiors appears to be the rule rather than the exception in ancient remains, regardless of whether they originate from polar or from tropical regions. The blackening of the lungs is termed *anthracosis*, and is as severe in some ancient tissues as it is in the lungs of ninetenth-century coal miners exposed to coal-laden air[4]. In its mild form it is probably quite harmless, but in more severe cases, particularly where combined with continual exposure to desert dusts, it may have led to *silicosis* and some impairment of the functioning of the lung. Modern studies of present-day communities living in the remote New Guinea highlands have suggested that lengthy exposure to smoke is detrimental to the health of the hut dweller.[5]

It is not possible to obtain samples of mummified lung tissue from ancient Britain, although there is evidence that in some periods the early British suffered from the effects of pollution indoors. The late Calvin Wells examined a large number of skulls from pre-industrial cemeteries in England and was particularly interested in recording the incidence of sinusitis, one of the diseases of the *maxillary sinus*[6] (Fig. 1.1). This sinus is one of the many air-filled spaces in the bones that surround the nose and give lightness to the skull. The spaces which open into the nose are lined with a continuous nasal membrane. This mucous membrane can become inflamed after the spread of infection from the nose or teeth. Repeated infection can destroy the cilia of the mucosal lining, allowing mucus to accumulate within the sinus instead of draining out through the opening into the nose. This condition often causes considerable discomfort to the sufferer, particularly in the winter months. Prolonged infection affects the bone at the floor of the sinus, which becomes roughened, or in severe cases pitted, as seen in Fig. 1.1.

The floor of the sinus of skulls from archaeological sites can be examined with modern medical instruments, such as the anthroscope. This is an optical device which allows photographs to be taken of the interiors of cavities with only small openings. It is thus possible to determine whether the individual had suffered from sinusitis. The results of examining just

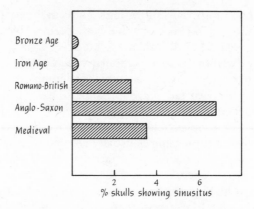

FIGURE 1.2 Sinusitis frequency in Britain through the ages

under 4300 skulls from various periods is given in Fig. 1.2. The low frequency of sinusitis in early times is striking in comparison with a high incidence in the Anglo-Saxon period.[7] It could be that the rather narrow Anglo-Saxon facial structure was more prone to the infection than faces of earlier periods, but it is quite possible that the disease was aggravated by environmental factors.

Dry dusty or cold damp extremes in climate seem to give rise to a high incidence of sinusitis. Particularly important among contributory environmental factors is housing, where inadequate ventilation can trap the smoke from hearths. It is probable that Anglo-Saxons suffered because their huts were not as well ventilated as earlier Iron Age ones, and by Romano-British times much of the cooking was done outdoors. There are also climatic considerations: the climate of the sixth and seventh centuries and maybe even the two that followed was rather cold and damp. Such weather conditions could have increased the frequency of sinusitis, but it would also have ensured that more time was spent indoors, so the degree of ventilation in huts could well be a particularly important factor in contributing to the high incidence of the disease in the early Anglo-Saxon period.

The problems of ventilation were partly solved by the development of the chimney, although it was probably not common in any but noble households much before the late sixteenth century. Smoky chimneys have continued to be a problem up to the present. Treatises on the subject have been numerous – many famous people such as Gilbert White, Benjamin Franklin, Count Rumford and Dr Arnott have all grappled with the difficulties of ridding our houses of smoke, a problem which Shakespeare says is as bad as a tired horse or a railing wife. The construction of houses has always been associated with ritual and superstition. Charms to cure smoky houses were probably frequent in early times and hints of them still remain in the present century, although by now the notion is merely poetic.[8]

FIGURE 1.3 Chimneys, even where they existed, might not have cleared the tops of medieval buildings.

Early urban pollution

It is evident from the previous section that even in the earliest times, combustion played a key role in the generation of air pollution. In the absence of urban areas with large conglomerations of people, air pollution must have been restricted to the contamination of interiors as suggested for the Anglo-Saxon period, but even in classical times big cities were likely to have had their own pollution problems. Classical writers seem to offer some support for this idea. The poet Horace mentions the blackening of buildings in Rome, and in many early cities this really must have been a problem. Seneca, Emperor Nero's tutor, suffered life-long ill-health and

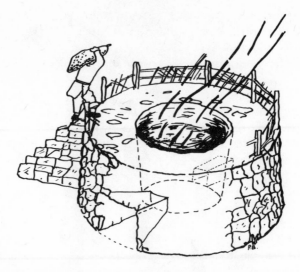

FIGURE 1.4 The lime kiln was a major source of air pollution in the
Middle Ages

was frequently advised by his physician to leave Rome. He wrote in a letter
to Lucilius (*epistolae Morales CIV*) about the year AD 61 that no sooner did
he leave Rome's oppressive fumes and culinary odours than he felt better.[9]

Ancient cities were small and inevitably suffered from overcrowding.
The cities needed to be small and densely populated to ease the problems of
defence and simplify movement of people and carriage of goods within the
urban boundaries. Under such crowded conditions smoke from small
forges or hearths must frequently have been emitted at a low level and
allowed to drift onto neighbouring houses. In Europe medieval cities often
became dominated by tall buildings. The streets between these densely
packed dwellings would have formed canyons, likely to trap smoke and
fumes. At first there was little concern over the efficiency of early chimneys,
which were often no more than a hole in the roof. This would have been
true even for semi-industrial establishments, as trades were often carried
out in rooms of an ordinary house after only the slightest of modifications.

In medieval London many of the small-scale industries must have
required a source of heat: ovens for bakers, kilns for brickmakers and
furnaces for metal workers. Even so the city's industrial requirement for
fuel must have been smaller than its domestic needs, which in winter would
be large. In less-developed societies today, the wood fuel consumption is
generally placed at about a tonne per year for each person. Fuel usage for
'cottage' industries such as bakers, dyers and candlemakers would only be a
little larger than this. To find large-scale use of fuel in medieval times, it is
necessary to look at industries concerned with the production of building

materials: pottery, tile, glass, iron, steel and lime – and even some of these would be rather small. Iron production from an individual forge might easily amount to just a few tonnes per year. Iron making, like many other early industries, would probably have been situated in the forest near a source of wood and well away from urban populations likely to complain of pollution. The products of the forest industries were compact enough to be transported with comparative ease. However, this was not true of the lime industry. Here the quantities involved were often very large as lime was extremely important to medieval society, being used in mortar and for agricultural application.

Lime was produced by heating limestone (calcium carbonate) to a high temperature in a kiln (Fig. 1.4). This drove off the carbon dioxide to produce lime (calcium oxide). When mixed with water to form a cement it was converted to slaked lime (calcium hydroxide). Traditionally limestone was burnt with oak brushwood. In requisition orders, such as those issued by Henry III in the building operations at Westminster in 1253, oak brushwood was specified as the fuel. It came by the boatload to the kilns at the building site. Change was so rapid that only eleven years later we find that a similar requisition order specifies a different fuel:

> To the Sheriffs of London   23 July 1264
> Contrabreve to purvey for the King in the City of London without delay and without fail a boatload of sea-coal and four millstones for the King's mills in Windsor Castle and convey them thither by water for delivery to the constable of the castle.[10]

Sea-coal or *carbonem marus* seems to have been so called because it was brought by sea to coastal centres in England during the thirteenth century. It must have appeared in London by early that century, since by 1228 there was a street in London called Sacoles Lane.[11] Its approximate location is still evident to the pedestrian in London today – both Old Seacoal Lane and Seacoal Lane are to be found near Ludgate Circus. There is much argument as to the reasons for the association of that area of London with sea-coal in the thirteenth century. It is often presumed that sea-coal was brought up the Fleet River and unloaded at Seacoal Lane and that much of the fuel was burnt in nearby Limeburners Lane. This is an attractive notion as the area near Ludgate had been associated with kilns since Roman times. While this idea receives support from the Elizabethan antiquarian John Stowe,[12] there is much evidence that coal was unloaded elsewhere. When Robert le Portour drowned while unloading coal in 1236, he did so in the waters of the Thames and not the Fleet.[13] The *Liber Albus*, which was compiled from earlier documents by John Carpenter during the lifetime of Mayor Richard Whittington, says the duty paid on coal at Billingsgate was to be one farthing for every two quarters of coal.[14] It would appear likely that the coal factors in the thirteenth century conducted their business with the captains of the

FIGURE 1.5 Eleanor of Provence, a forthright and unpopular queen, complained about air pollution in 1257

colliery vessels in an open space known as *Romeland* near Billingsgate.[15] It is easier to support the argument that Seacoal Lane was a residential area for the coal merchants who lived in London, because a William of Plessey lived there in 1253.[16] Despite the difficulties in establishing the way in which Seacoal Lane came to be named it does signify a very early beginning to the importation of coal into London.

The earliest documented air pollution incident in England occurred not in London but in Nottingham. It is interesting to consider this case and to note the earliest reactions towards the new fuel. In the 1250s Henry III initiated repairs to Nottingham Castle. The sheriff of the castle, Robert le Vavasser, who was responsible for the work, managed the operations very badly.[17] A timely death allowed him to escape from the consequences of gross mismanagement. Nevertheless Henry III sent the Abbot of Pershore and William de Walton to investigate. This and subsequent investigations showed the need for further repairs, which were probably under way when Queen Eleanor visited the castle late in the summer of 1257.[18] She found the air so full of the stench of sea-coal smoke that she was forced to leave for

Tutbury Castle to preserve her health. The fear of damage to health is notable in this early comment. Similar fears permeate almost all medieval complaints about air pollution.

This is so marked that it is necessary to ask why the sulphurous smell resulting from coal burning suggested to the medieval mind that the fumes were unhealthy. Certainly, foul smells had been long related to unhealthy airs. The Greeks gave the name *miasma* to the unhealthy odours that arose from swamps. We still use the word today but to mean a poisonous or infectious atmosphere. There were medieval examples of the fear of marshes; the site of Winchester Cathedral had to be moved because of the unhealthy and foul-smelling bog that had formed around it,[19] and the River Fleet in London was so malodorous that the monks at White Friars claimed that some of their brethren had died from the stench.[20] This type of popular perception of the origin of diseases may have allowed ready association between the smell of burning coal and the effects of air pollution on health.

Early attempts at control

In medieval London the pollution from coal burning was regarded as such a serious matter that a commission was set up to investigate the problem in 1285, and it was convened again in 1288 with firm instructions to find a solution.[21] The members of this commission remained fairly constant over a quarter-century, with the same names appearing in the documents. Perhaps officials such as John de Cobbeham or Ralph de Sandwich were England's earliest air pollution experts. Sadly the details of their deliberations have been lost. However, we do know that the outcome of a meeting in 1306 was a proclamation banning the use of sea-coal, but further legal notices issued only two weeks later suggest that the proclamation had been largely ignored.

It is popularly believed today that one of the early offenders against these air pollution laws was hung, tortured or decapitated in 1307,[22] although none of the writers who make the suggestion gives any primary reference for the incident. It is unlikely that the punishment would have been quite so severe even in such turbulent times. The proclamation of 1306 suggests that the offenders be punished by *grievous ransoms*.[23] One would guess that this meant fines, removal of the furnace and confiscation of tools, because these were the punishments usually enforced when kilns were unlawfully erected on roads.[24]

One cannot fail to be surprised at the smoke abatement activities of the early municipal government in London, when local government today appears to be so slow to control air pollution. This activity is particularly remarkable when the quality of the urban environment was so very poor in medieval times, with dung and rubbish frequently lying about in the streets. However, the administrators were concerned that towns should remain attractive to important visitors and untidy streets were frequently cleaned up before the meetings of Parliament, in order to encourage as many wealthy people as possible to come from their country retreats into the city. This is likely to be part of the explanation of municipal activity; some early documents relating to air pollution stress that the air pollution would cause offence to visiting nobles and prelates. It is a characteristic of visitors to cities, evident even today, that they are more likely to notice air pollution than are the people who have accustomed themselves to a polluted environment by living perpetually within it.

The strength of the reaction against air pollution may have been concentrated by the fact that it could be directed against rather small groups of people who were already disliked. These groups included the kinds of businessmen and petty industrialists who could influence the price of goods in medieval London. The butchers, bakers and brewers not only fouled the waterways but were also the subject of many complaints about pricing. The limeburners, who were blamed for early air pollution, were also often accused of unfairly raising prices. The fact that many colliers and some limeburners would have been Northerners cannot have enhanced their popularity!

A particularly interesting case which reflects further discrimination is one that was brought against the Knights Templar in 1306 by Henry Lacy, the Earl of Lincoln.[25] He accused them of blocking the river Fleet by constructing a watermill. The brethren of White Friars had complained of the same stench from the river. A commission was set up to investigate the cleaning of the Fleet in the summer of 1307. The legislators realized, even at this time, that the source of the offensive garbage that filled the river was the discharge of tanning waste and offal by the butchers and leatherworkers at Smithfield market. Only a few years before, Richard de Houndeslowe, a tanner, was brought before the mayor's court for corrupting the air to the danger of the Brethren of the Order of St Augustine. He swore he would no longer skin any carcases within the city or throw them into the ditches within or without.

Despite the fact that the problem arose from discharges into the river, the Knights Templar, as owners of the mill at the mouth of the Fleet, were blamed for the pollution (Fig. 1.6). No doubt their mill caused some blockage of the flow of the river, but it would seem likely that there was more at stake in this incident than water quality. The wealth of this religious order and their freedom from control by the secular authorities had made them unpopular. Persecution of the Knights Templar became more intense in 1307, because Philip IV of France wanted to extract money from them for his Flemish War. The religious order was dissolved by a papal bull issued at the Council of Vienne in 1312. By noting the extreme

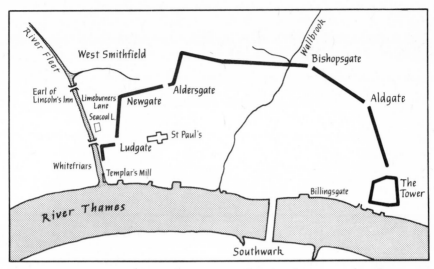

FIGURE 1.6 Map of London *c.* 1300 showing important locations mentioned in the text

unpopularity of this group of people it is easy to see this incident as an early case in which the stated environmental concern may simply have been a method of obscuring political objectives.

The problem of using an environmental problem as an instrument for political change is that this frequently does not lead to improved environmental quality. In the case in point, the waste from Smithfield and the privies along the Fleet continued to be ejected into the river. There were further problems from the 'infection of the air' and the abominable stench from the discharges of latrines and tanning works in 1355.[26] Pollution of the Fleet remained a problem until the time of Charles Dickens, five centuries after the expulsion of the Knights Templar.

While the courts and Parliament tried to control air pollution within medieval London, their influence was slight compared to the effects of the normal patterns of life typical for the period. This may be seen in the seasonal distribution of pollution incidents brought to the notice of the civic authorities (see Fig. 1.7(b),. If we assume that the number of complaints about air pollution is related to the frequency of the pollution problems, we can postulate that pollution in London at this time was very much a summer problem. A seasonal distribution of this type demonstrates that the usage of coal was not for domestic heating, since use for heating would be expected to show a winter peak rather like that seen today.

Complaints found in thirteenth-century documents point to the lime industry as being the main source of air pollution within medieval London. Such a conclusion would be consistent with the large amounts of coal burnt in lime production (thousands of tons) compared with only a ton or so used

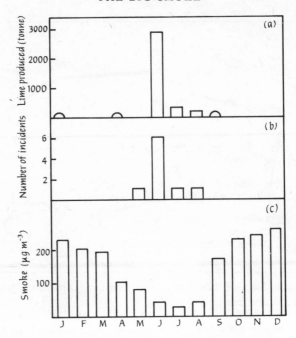

FIGURE 1.7 (a) Seasonal use of lime in thirteenth-century building operations compared with (b) the number of air pollution incidents and (c) modern concentrations of smoke in London's air

by a single forge in a year. Lime production in the thirteenth century shows the same seasonal patterns that are seen for complaints about air pollution in the capital (Fig. 1.7(a)). Admittedly the lime production levels used in the figure come from the records kept during the construction of Harlech Castle in Wales in 1286, but building operations in London must have been little different. The seasonal distribution observed in these figures for lime usage is as expected because construction work took place during the summer. When coal was required it had to be ordered from Newcastle early in spring. The building work closed down for the winter when shorter days and bad weather made outdoor construction difficult. In the London of the present century, when heating is responsible for a significant proportion of fuel consumption, a noticeable winter peak may be found in pollution levels (Fig. 1.7(c)).

Not all the litigation over air pollution concerned the major sources. Quite ordinary people seemed to bring complaints before the authorities. They were quick to realize that pollution could effect property values. For instance, in the fourteenth century a case was brought before the London Assize of Nuisance which illustrates this:

> Thomas Yonge and Alice his wife complain . . . the chimney is lower by 12ft than it should be and the blows of the sledge-hammers when the

FIGURE 1.8 Blacksmiths were blamed for much early environmental pollution

great pieces of iron called 'Osmond' are being wrought into 'brestplates', 'quysers', 'jambers' and other pieces of armour, shake the stone and earthen party-walls of the house so that they are in danger of collapsing, and disturb the people and their servants day and night, and spoil the wine and ale in their cellar, and the stench of the smoke from the sea-coal used in the forge, penetrates their hall and chambers, so that whereas formerly they could let the premises for 10 marks [£6 13s 4d] a year, they are now worth only 40s.[27]

It may be that this case was rather like the one concerning the Knights Templar and that the charge of environmental pollution was simply a ruse to get rid of noisy neighbours.

The defence offered by the armourers, in this case, is interesting, because it is similar to that which would be offered by many small industries today. They claimed that they were honest tradesmen and should be free to carry out their trade anywhere in the city, adapting their premises to suit the work. They rejected the complaint of nuisance on the basis that the forge they used had long been on the site and the rooms affected by smoke had been built more recently.[28]

Despite the fact that blacksmiths (Fig. 1.8) used only small quantities of coal, there was considerable reaction against the environmental nuisance they caused. Much of this derived from the noise, dirt and long hours worked, as recorded in an interesting medieval satire:

> Sooty smoked smiths, smattered with smoke,
> Drive me to death with the din of their efforts,

Such noise at night a man never heard,
With the knaves shouting and the clatter of blows!
The crooked connivers cry 'Coal! Coal!'
And blow their bellows till their brains near burst,
'Huff, puff', says one and 'Haff, paff' the other.
They spit and sprawl and spin tall stories,
They gnaw and gnash and groan together,
Are kept hot heaving hard heavy hammers.
Their aprons are of bull hide,
Their shanks are sheathed against the sparks.
Huge hammers are handled hard,
Strong strokes struck on steel stock.
'Lus, bus, las, das', tapping in turn.
Oh the Devil end this dreadful din.
The master lengthens pieces of iron,
Twining and twisting them with terrible twanging,
'Tik, tak, hic, hac, tiket, taket, tik, tak,
Lus, bus, lus, das'. Such a life they lead,
Christ punish these horse-shoe benders,
Who cake our clothes and ruin our night's sleep.[29]

The control of air pollution

There is much uncertainty about the success of early attempts to control air pollution in London. The outright ban that was placed on the use of sea-coal by local government was virtually the only effective response. It is possible that they may also have encouraged the construction of efficient or tall chimneys. And after all we find in the case from *Assize of Nuisance* just discussed that the plaintiffs argued that the chimney of the armourers was not as high as good practice required. Today high-stack policy is one of the main ways in which local pollution is avoided close to large power stations. High chimneys not only carry the pollutants high above our heads, but they can also take advantage of the stronger winds aloft. Under stagnant conditions stable layers of air, known as temperature inversions, can trap pollution near ground level. A tall chimney stack can sometimes penetrate through the stable layer of air, but with medieval chimneys the main consideration would have been to ensure that the smoke was removed from the interiors and emitted at sufficient height to clear the neighbouring houses.

A number of other approaches to air pollution control were explored in medieval England. One idea was to place limits on the use of coal. Restriction on the times when furnaces could be used was suggested by one group of London blacksmiths in the thirteenth century.[30] In late

FIGURE 1.9 King Edward I, England's Justinian, initiated some of the earliest air pollution legislation

medieval Beverley it seems that a zoning ordinance was set up to ensure that brick kilns were kept at some distance from the town because of the damage caused to fruit trees.[31]

It is more than likely that the legislation of the late thirteenth and early fourteenth century had little effect. The proclamations of 1307 may have been partially successful because they were specifically promulgated to control lime kilns burning sea-coal. The complaints about smoke that arise three years after this legislation do not specify a fuel.[32] They simply state that the kilns should be moved because the air was corrupted to the inconvenience and peril of men in nearby parts of London. If these complaints of 1310 are not concerned with air pollution from the use of coal this is rather interesting. It may mean that the legislation of 1307 had been successful, for a short time at least, in preventing the use of sea-coal in kilns. However, it cannot have been so successful against the use of coal in furnaces, because it is apparent a few years later that sea-coal was being used for forging. Perhaps the small amounts of coal used in forging were exempt from the regulations. Whatever the regulations, the dislike of blacksmiths suggests that the public were not willing to make special concessions even if the legislators were.

Even if we can accept the notion that for a short time there was some control on the use of coal in lime burning in the very early part of the fourteenth century, these controls cannot have been long lived. In 1329 there is evidence that lime was being burnt with coal again since in that year a crooked collier from Northumberland, Hugh Hencham, was brought before the courts for 'fixing' the prices of lime in London.[33]

From this point until the mid-fourteenth century there appears to be a gap in the records of complaints about pollution in London. Despite the gap, the arguments above imply that we cannot conclude that air pollution was no longer an important problem. It may simply be that the events were no longer recorded. However, circumstances in London may have changed slightly. Wood for fuel may have been more plentiful, or perhaps the citizens grew accustomed to the smell of coal smoke, even that in association with lime burning. Until Elizabethan times lime burning remained one of the major uses of sea-coal and Shakespeare complains of the reek from lime kilns.

Fuel shortages

An important point we have not considered is: just why didn't the air pollution problem appear in London until the late thirteenth century? After all, coal had been available in the city from the very beginning of that century. It seems that initially the fossil fuel found few uses, but the gradual increase in the price of wood throughout the latter half of the thirteenth century promoted the use of coal as an industrial fuel. In contrast to wood, coal remained at a relatively constant price through much of the thirteenth century as it was shipped south simply as ballast. The financial burden imposed by soaring wood prices may have been strongly felt by the lime-burners, as the price of lime, like that of many medieval commodities, was fixed by ancient ordinances. These took little account of inflation in the prices of the raw materials.

The increase in the price of wood fuel may have resulted from shortages which arose from the population growth of England under the Norman kings. The climate of Europe was at its warm optimum, so extensive areas of land were brought under cultivation. In the twelfth century Fitzstephen wrote that plentiful forests abounded in the counties adjacent to London, but as with the rest of England the population of the city grew and so did the pressures of clearance and utilization. The forests around London were gradually depleted.

In general it is observed that the gradual elimination of forests at the urban perimeter ensures that one early result of city growth is a change from wood fuel to charcoal which is more easily transported. The manufacture of charcoal is a rather smoky process, but production takes place in the forest. Charcoal is a smokeless fuel so its use lowers urban air pollution. The abundance of charcoal at the production site, in the forests, means that industries such as glass and iron-making, which utilize it, develop alongside. Thus in early times the pollution from forest-based industries was of little importance to the urban dweller.

The use of charcoal offered only a temporary solution to London's fuel requirements. The increasing population put a great strain on the fuel resources and such pressures almost inevitably led to a growth in the air pollution problem. Despite the fact that many industries were located in the forests of the Weald, increasing amounts of fuel were needed by the growing city. Attendant price increases encouraged the adoption of newer, cheaper fuels such as coal. The earliest air pollution incidents in London, at the end of the thirteenth century, came at a time when its population was at a peak.

The eleventh and twelfth centuries had been periods of rapid population growth in Europe, initially spurred on by improved agriculture and relatively efficient feudal government. The impact of this expansion in England is illustrated by the fact that little land remained to be cleared by the thirteenth century. The population was still growing,[34] and plots of arable land became fragmented, so much so that after 1250 living standards began to drop as population growth began to outstrip food production. Food shortages were common throughout the fourteenth century. It seems as if the feudal society had been too successful for its own good. Stagnation followed: wages and food prices increased, malnutrition became more common and The Black Death ravaged an already weakened population between 1348 and 1351.

As much as a quarter of the population of England died in the initial outbreak of the disease; attacks were often repeated and the famine that ensued worsened the level of health. It is estimated that the plague may have killed 25 million people from a European population of 80 million between 1348 and 1351.[35] In England further serious outbreaks of plague occurred in 1360–1, 1369 and 1374. The cities were often worse off than the countryside, as plague was particularly virulent where population was dense and the standard of housing low. This unprecedented decrease in the population during the fourteenth century meant much land was no longer cultivated. Some became pasture, some grew into brushwood and finally forest. The standard of living then rose and governemnt became more centralized. Firewood, once a prohibited export, began to be exported again by the mid-fourteenth century. So the first fuel crisis was over and wood was more plentiful, at least for a short time.[36]

The lesson of this early environmental crisis is an important one because the sequence of events has repeated itself more than once since then: rapid population growth, urbanization or increases in population density, fuel shortages and changes in the pattern of fuel use. Unfamiliar fuels are liable to be more polluting than their predecessors or at least are perceived to be so. All these elements are to be found in the changes that occurred in thirteenth-century London. The same key elements may be found in the problems of the London of the seventeenth century and of more recent times.

Notes

1. Henry Ford might almost be credited with this seminal observation, the corollary of which, 'History is rubbish', has been a central theme of much archaeological research: see Keene, D. (1982) 'Rubbish in medieval towns', *Environmental Archaeology in the Urban Context*, Council for British Archaeology Research Report 43, or Moore, P. D. (1981) 'Life seen from a medieval latrine', *Nature*, 294, 614. Two books sympathetic to this view of history are: Zinsser, H. (1935) *Rats, Lice and History*, Atlantic Monthly Press, Boston and Cloudsley-Thompson, J. L. (1977) *Insects and History*, Weidenfeld & Nicolson, London.

2. Two works which deal with social and ethical aspects of environmental thought are: Barbour, I. G. (ed.) (1973) *Western Man and Environmental Ethics*, Addison-Wesley, Reading, MA, and Tuan, Y. F. (1975) *Topophilia: A Study of Environmental Perception, Attitudes and Values*, Prentice-Hall, Englewood Cliffs, NJ.

3. For example, the American Indian wigwam as seen in Vivian, J. (1976) *Wood Heat*, Rodale Press, Emmaus, PA, 24–5.

4. Anthracosis is the rule rather than the exception in mummified lung tissue, from Alaska to Peru. The latter is particularly interesting, silver ore in the lung suggesting occupational damage through work in the mines. See Bothwell, D. R., Sandison, A. T. and Gray, P. H. K. (1959) 'Human biological observations on a Guanche mummy with anthracosis', *Amer. J. Phys. Anthrop.*, 30, 333; Zimmerman, M. R., Yeatman, G. W., Sprinz, H. and Titterington, W. P. (1971) 'Examination of Aleutian mummy', *Bull. N.Y. Acad. Med.*, 47, 80–103 – this mummy, like so many, had anthracosis; Cockburn, A., Barraco, R. A., Reyman, T. A. and Peck, W. H. (1975) 'Autopsy of an Egyptian mummy', *Science*, 187, 1155; Zimmerman, M. R. and Smith, G. S. (1975) 'A probable case of accidental inhumation of 1600 years ago', *Bull. N.Y. Acad. Med.*, 51, 828–37; Gerstzen, E., Munizaga, J. and Klurfeld, D. M. (1976) 'Diaphragmatic hernia of the stomach in a Peruvian mummy', *Bull. N.Y. Acad. Med.*, 52, 601–4; Reymen, T. A., Zimmerman, M. R. and Lewin, P. K. (1977) 'Autopsy of an Egyptian mummy (Nakht-Roml). S. Histopathologic investigation', *Can. Med. Assoc. J.*, 117, 7–8.

5. Cleary, G. J. and Blackburn, C. R. (1968) 'Air pollution in native huts in the Highlands of New Guinea', *Archs Envi. Health* 17, 785–94; Master, K. M. (1974) 'Air pollution in New Guinea – cause of pulmonary disease among stone-age natives in the Highlands', *JAMA* 228, 1653–5.

6. Wells, C. (1977) 'Diseases of the maxillary sinus in antiquity', *Medical and Biological Illustration* 27, 173–8; before his death in 1978 Calvin Wells told me that all the additional work since the above paper was written had only strengthened its conclusions.

7. An interesting comparison with the English figures comes from the examination of nearly 500 Indians massacred in South Dakota in the late fifteenth century. Only five cases of sinusitis were found, suggesting a frequency of about 1 per cent. Grey, J. B. 'The post mortem at Crow Creek', *Paleopathology Newsletter*.

8. Morley, C. D. (1921) *Chimney Smoke*, George H. Doran, New York. This wonderfully titled book offers the following piece of doggerel:

Dedication for a fireplace
A Charm

O wood, burn bright; O flame, be quick
O smoke, draw cleanly up the flue –
My lady chose your every brick
And sets her dearest hopes on you.

Logs cannot burn, nor tea be sweet
Nor white bread turn to crispy toast,
Until the charm be made complete
By love, to lay the sooty ghost.

9. Lord Dunsany (1947) *The Odes of Horace*, Heinemann, London emphasizes that Horace is an essentially rural poet and objects to the Roman equivalent of urban expansion. For Seneca, see Gummere, R. M. (ed.) (1971) *Ad Lucilium Epistolae Morales III*, Heinemann, London.

10. *Cal. Lib. Rolls*, 37 HIII 9; *Cal. Lib. Rolls*, 48 HIII.

11. *Pipe Roll*, 12 HIII.

12. Stowe, J. (1603) *A Survey of London.*

13. *London Eyre of 1244*, Lond. Rec. Soc. (1970).

14. Riley, H. Y. (ed.), (1861) *Liber Albus.*

15. See 'Coal', in Kent, W. (1951) *An Encyclopaedia of London,* Dent, London. However, there were a number of Romelands in the city, notably at Queenhithe. It is possible that Romeland was a name given to an open space near a dock where cargoes could be discharged (Dugdale, W. (1693) *Monasticon Anglicanum*, J. Wright, London).

16. *Cartae Antiquae*, Chancery, L., no. 20 *in dorso*.

17. *Cal. Pat. Rolls*, 39 HIII m 15d.

18. *Annales de Dunstaplia Rerum Britannicarum Medii Aevii Scriptores, Annales Monastici III*, Longman, Green, Reade & Dyer (1866). A disturbingly inaccurate account of this incident occurs in the historical chapter of a recent text on air pollution, where an opening sentence reads: 'Air pollution associated with burning wood in Tutbury Castle in Nottingham was considered "unendurable" by Eleanor of Aquitaine, the wife of King Henry II of England and caused her to move in the year 1157.' This sentence contains no less than six factual errors; see Brimblecombe, P. (1952) 'An anecdotal history of air pollution', *Environmental Education and Information*, 2, 97–105:

(i) the year was 1257; (ii) the king was Henry III; (iii) the queen was Eleanor of Provence; (iv) the fuel was coal (i.e. sea-coal or 'carbonum maris'); (v) Tutbury Castle is in Staffordshire; (vi) the pollution occurred in Nottingham Castle, not Tutbury Castle.

The quality of air at Tutbury Castle did not remain high over the centuries. Mary, Queen of Scots, complained bitterly of the stench of privies that pervaded the rooms while she was there in 1585; see Strickland, A. (ed.) (1844) *Letters of Mary, Queen of Scots*, Henry Colburn, London, I, 163.

19. 'Excavation near Winchester Cathedral', Dugdale, W. (1693) *Monasticon Anglicanum*, J. Wright, London.

20. *Rot. Parl.* I, 61b and I, 200.

21. *Cal. Pat. Rolls*, 13 EdI m18d; *Cal. Pat. Rolls*, 16 EdI m12; *Cal. Close Rolls*, 35 EdI m6d and m7d; *Cal. Pat. Rolls*, 35 EdI m5d; *Cal. Close Rolls*, 34 EdII m23d. The first two references are to the commissions of 1285 and 1288 while the others are also important medieval proclamations. There are frequent secondary references in books on air pollution to a law passed in 1273 (e.g. Magill, P. L., Holden, F. R. and Ackley, C. (1956) *Air Pollution Handbook*, McGraw-Hill, New York. I can as yet find no trace of a primary reference to this event.

22. Maus, O. and Chubb, L. W. (1910–11) 'Smoke', in *The Encyclopaedia Britannica*, 11th edn, vol. XXV, 275–7; Chambers, L. A.; 'Classification and extent of air pollution problems', in Stern, A. C. (ed.) (1968) *Air Pollution*, vol. I, Academic Press, New York; and Perkins, H. C. (1975) *Air Pollution*, McGraw-Hill, New York.

23. Ransom as a sum of money paid to redeem a person guilty of a great crime is defined in *Stroud's Judicial Dictionary*, 4th edn (1974), Sweet & Maxwell, London.

24. Relevant to the dismantling of kilns see Chew, H. M. and Weinbaum, M. (1970) *The London Eyre of 1244*, London Record Society, cases 350, 351 and 470.

25. See Honeybourne, M. 'The Fleet and its neighbourhood in early medieval times', *London Topographical Record*, 19, 13–87; *Cal. Pat. Rolls*, 35 EdI 9d; *Cal. of Plea & Mem. Rolls*, Roll F m7.

26. *Select Pleas in the Court of the King's Bench, I*, Selden Society, London (1936), 55.

27. *London Assize of Nuisance, 1301–1431*, London Record Society (1973).

28. A rather similar incident arises in the life of the painter Sandro Botticelli. He was being deafened by noisy neighbours whose work shook the whole house. They insisted that they could do what they wished in their own home. Botticelli's house was higher than the neighbours' so he balanced an enormous boulder atop his house which threatened to fall at the least vibration, claiming that he too was free to do as he wished in his own home. Vasari, G. (1965) *Lives of the Artists*, Penguin, Harmondsworth.

29. Wright, T. (1845) *Reliquiae Antiquae*, London vol. I, 240. Also found in Davies, R. T. (1963) *Medieval English Lyrics*, Faber & Faber, London, author's translation.

30. *Cal. of Early Mayor's Court Rolls*, Roll B m5.

31. Leach, F. (1900) 'Beverley town documents', *Seldon Soc.*, 14.

32. *Cal. Close Rolls* 4 Ed II m23d.

33. *Cal. Letter Book of London E*, Fo. cxcvii, cxxcviib.

34. Trevelyan, G. M. (1960) *Illustrated English Social History*, vol. I, Longman, London.
35. Cipolla, C. M. (1976) *Before the Industrial Revolution*, Methuen, London.
36. My interpretation of the broader impacts of the Black Death can be partly supported by recent works such as Gottfried, R. S. (1983) *The Black Death*, Free Press, New York. Here the wide range of changes that occurred are treated; in particular, some of the benefits that came with declining population. Although this reference does not treat the pollution aspect specifically, the epilogue, interestingly enough, is titled 'Europe's Environmental Crisis'.

2

The rise of coal

The importance of coal in the earliest documented air pollution episodes in the city of London was discussed in the previous chapter. The use of coal in London in the Middle Ages was probably never very large. However, it did grow and by the end of the reign of Elizabeth I (1558–1603) it had reached 50,000 tonnes per year. The fourteenth, fifteenth and sixteenth centuries saw little real change in the physical size of London. While it is true that at the end of this period there was a considerable increase in population, the actual change in area of the city was comparatively small. However this did mean that, particularly within the city walls, there was an increase in population density which accounted for many of the new problems the citizens of London had to face. The increasing density of human activity and a consolidation of the built environment would have caused marked changes in the atmosphere and climate of the city. This chapter will follow those changes. It was a period through which air pollution seems to have been relatively unimportant, but the changes under way led to the problems that became evident at the end of the sixteenth century.

Climate and the city

Air pollution is of course an atmospheric phenomenon. This is obvious but often we neglect the way in which air pollution is linked to climate. It is necessary to consider climate, climate change and the microclimate of the city itself if we are to understand the air pollution history of late medieval London.

The climate of Britain, despite its critics, is a remarkably moderate one. It is not at all unsuitable for human comfort, protected as it is from extreme fluctuations in temperatures. Very hot climates, where the body finds it difficult to release metabolic heat, lower general vitality and increase the incidence of infectious diseases. Cooler climates may be stimulating, but cold climates enhance susceptibility to respiratory diseases, especially when the low temperatures mean that people have to spend much of their life crowded into poorly ventilated, heated rooms. The temperature on summer afternoons in southern England is about 20°C; while this is optimum for human comfort it is somewhat low for efficient growth of certain crops such as cereals, so their yields were often meagre in the past. The winters in England are mild for such latitudes, owing to the transfer of warm water to the western coasts by the Gulf Stream and North Atlantic Drift. In winter freezing temperatures are generally restricted to the night, but at times there are exceptionally cold spells. Such situations arise when north-easterly winds bring air down from the Arctic, and cold weather may persist for weeks if the meteorological conditions become stable or 'blocked'. However, it is not always wintry weather that may result from blocking conditions. Long anticyclonic spells in the summer are also very much a part of English weather.

Tourists are quick to notice that there are no dry seasons in the climate of the British Isles. The annual cycle of rainfall in London shows rather small seasonal differences,[1] but it is not simply the amount of rainfall that is important, as communities soon adapt to the level experienced in their particular region. It is variation about this mean which causes stress, showing itself in a declining water supply to an urban population or in floods, both of which may result in poor agricultural yields. While it is true that the variation is at its largest in southern England, only in one year in twenty will rainfall be expected to deviate from the mean by more than 40 per cent.[2] On a seasonal level variations can be larger, and of course this can be more critical in terms of crop yields.

Wind can become a particularly destructive meteorological element but its effect is restricted to extreme events such as gales. It is interesting to note that extreme gusts are rarer in the area around London than anywhere else in England.[3] Windy conditions are useful in drying damp and possibly disease-ridden parts of cities. While it is possible to describe the climate of England in the very general terms used above, it is important to realize that climate is not static. The seasonal variations have already been mentioned, along with the notion that some year-to-year variability in the weather is found, with dry years and wet years or cold and warm ones. Beyond these inter annual variations there are long-term trends in climate. The eleventh and twelfth centuries were marked by warm conditions and are often known as the Medieval Warm Epoch, which was characterized by the development of English vineyards. This warm period was followed by a cooler one. Between the fifteenth and eighteenth or even nineteenth centuries the mean temperatures in England were probably a degree or two less than those of the present day: this period has been called the Little Ice Age. Throughout this period the climate was different in a number of ways: extreme weather events such as storms became more frequent, the rainfall amounts underwent wide oscillations over 50-year periods, and some summers may have been extremely wet, although winter rainfall seems to have been less affected.[4]

In the countryside storms can easily damage crops. In cities the changes

brought about by urbanization allow heavy falls of rain to have other
effects. The Roman wall that surrounded London had an enormous culvert
to allow the Wallbrook to flow through the city (see Fig. 1.6, p. 11). In the
post-Roman period this culvert was so frequently blocked that in spite of
some attempts to keep it clear, a great area of land behind the wall flooded,
forming a marsh.[5] The problem of storm drainage, however, is more
complex than simply blocked watercourses. It is often hard to appreciate
the enormous amount of drainage that a city needs. The surface onto which
rain falls in the city is more 'waterproof' than in the country. Building
materials have a low storage capacity, which means they don't become very
wet and cannot hold such large quantities of water as vegetation or soil.
When a storm hits a city the water appears directly as run-off. The
discharge through drains closely follows the input from the storm, the flow
off the urban surface reaching a peak shortly after maximum rainfall
intensity. Flow from a surface with vegetation cover, on the other hand,
can occur much later and have a much lower discharge. This is the reason
why storm sewers have to be exceptionally large in urban areas. In the
absence of an effective system of drainage, flooding may be extensive after
storms. In early cities, which had no adequate storm drainage, floods
caused considerable hardship in low-lying areas immediately after large
storms.

The surface flow that occurs during storm drainage is often highly
polluted and in past times was no less so, picking up contamination from
rubbish and foul sewage, thence contaminating the drinking water. There
are frequent accounts of this problem in medieval documents, but a more
interesting one appears in Swift's *Description of a City Shower* (1711):[6]

> Now from all parts the swelling kennels [drains] flow,
> And bear their trophies with them as they go:
> Filth of all hues and odours seem to tell
> What street they sail'd from, by their sight and smell . . .
> Sweepings from butchers' stalls, dung, guts and blood,
> Drown'd puppies, stinking sprats, all drench'd in mud,
> Dead cats and turnip-tops, come tumbling down the flood.

High rainfall made the medieval roads treacherous and as we shall see later
this had a profound effect on the transport of fuel into the city. In times of
drought another problem may have arisen. The streams and channels, used
as open sewers in early cities, became blocked and polluted when the flow
of water was low. The stench from these foul watercourses was a subject of
frequent complaint, as we saw in the case against the Knights Templar in
the previous chapter.

The construction of extensive urban areas changes the local climate. The
best known of these changes is an increase in temperature in the built-up
area, which is often called the 'urban heat island'. This does not arise simply

from the fact that industrial activity and domestic heating processes increase the air temperature. There is a range of factors which lead to a rise in temperature. The urban surface may be darker in colour, thus absorbing more heat, and the vertical surfaces tend to capture sunlight effectively when it is at low angles. The materials from which the city is built have a high storage capacity for heat which means that they can remain warm well into the night. Walls and buildings can create sheltered areas, so that wind chill is much decreased. However, this is not always true because around the sharp corners of buildings the wind can be very gusty. At night re-radiation from the vertical walls can be considerably greater than in open country. Finally it is possible for the smoke particles suspended in the air above the city to keep the city warm through a blanketing effect which traps heat radiated from the ground.[7]

Naturally there are no medieval measurements of the temperature of London to give us an idea of the magnitude of this urban heat island. Some archaeologists have noted the presence of warmth-loving beetles in English cities of the Little Ice Age and have attributed their presence to higher temperatures in the urban environment compared to the rural one, though the presence of these beetles in cities must have also been encouraged by the increased amount and variety of food to be found there. The urban heat island may have some important consequences for urban flora and fauna. Birds which are small and lose heat rapidly are able to take advantage of the elevated temperatures of the city in winter. This is particularly convenient for birds that have made use of other aspects of the urban environment: the buildings in cities often provide numerous nesting areas, while the inhabitants and their activities provide a ready source of food.[8]

Plants can also benefit from the extra warmth offered by the urban environment. Flowering of some plants in cities can occur earlier than in rural areas and the growing season may be lengthened by as much as five weeks. Of particular significance in the British Isles is the persistence of vineyards well into the Little Ice Age. Manley[9] has suggested that some increase in temperature occurs in carefully tended, sheltered gardens, which can have a mean annual temperature several degrees higher than temperatures in exposed surroundings. The gain in temperature arises chiefly from the restriction of the re-radiation of heat. However, crops in such gardens no doubt also benefit from the calm conditions and higher humidity which would lead to lower loss of water by transpiration. Nocturnal air temperatures in such sheltered areas tend to be lower than in exposed locations, but if the soil is well irrigated and compressed by frequent trampling during cultivation of the vines, the temperature at night can actually be a degree or more warmer than that of the unsheltered surroundings.[10]

The changes that occur through the presence of an urban environment

TABLE 2.1 Changes in climate expected with urbanization

Temperature	rise of some degrees
Fog frequency	increase, particularly in winter
Thunderstorm frequency	possible increase
Cloudiness	possible increase
Wind	change in speed and distribution
Sunshine	decrease, particularly in the ultraviolet
Snow cover	changes in distribution

are listed in Table 2.1. The changes are actually rather small in magnitude and not easily perceived by non-instrumental observations, although travelling directly from the country into the city would make them more noticeable. The earlier flowering dates of urban plants could probably be observed quite readily. In contrast to these rather small climatic changes the changes in composition of the atmosphere in cities can be very large. It is this difference in magnitude that has resulted in so much of our concern today being directed at the pollution of the urban atmosphere, rather than the climatic change that results from urbanization.

It is against this meteorological background that Tudor London grew. Long-term climatic changes meant that the temperate oceanic climate was entering a cooling phase. The population of England was beginning to recover slowly after the ravages of numerous plagues, despite famines that occurred on an almost annual basis. The recovery in population brought with it new arts and industry. The Renaissance was perhaps slow in reaching this remote and insular nation, but its impact was not weakened by distance. The growth of the city of London paralleled the intellectual renaissance and London's population entered five figures.

The transport of fuel

The amount of coal imported into London during the thirteenth century can have increased only marginally in the centuries that followed. This can be seen by the fact that the number of coal-meters who measured and taxed the sea-coal brought through the wharves at Billingsgate did not increase from the time of Edward II to Queen Elizabeth's reign (1307–1558). Despite the slow increase in the quantity of coal imported by the capital, the fuel was becoming a more important commodity. In *The Chronicle of Grey Friars*[11] 'a gret derth for wode and colles' is noted for the year 1543. When the preferred fuels, wood and charcoal, were scarce, coal was used in their place.[12]

Shortage of wood as a fuel in growing cities is not uncommon. At first the area around the city becomes deforested and turned over to agricultural

FIGURE 2.1 Wood being loaded onto carts for transport into the city

land which yields a higher profit to the landowner. This situation has been familiar in developing countries in recent times. It means that the fuel must be transported into the city over increasing distances; as these lengthen the cost of transport becomes a major part of the fuel price. The fuel used often switches to a conveniently packaged form such as charcoal, which is more easily transported into the city. For a given weight charcoal gives more heat than wood, and as it lacks branches it is easier to pack into carts. Fuel shortages in Tudor times were not nationwide but seem to have been restricted to certain regions. Wood was particularly scarce in the depleted forests close to large cities.

The other advantage of charcoal, in a city where chimneys are poorly designed or absent, is that it is a smokeless fuel. Smokeless fuels are often remarkably successful in reducing pollution, but their production may itself cause severe local pollution problems. This is true even today when the production of smokeless fuels such as Phurnacite has been seen as relocating the very environmental problems it attempts to cure.[13] The sites for charcoal production would generally have been far off in the forest where they would be free from complaint. However, mercantile pressures no doubt encouraged charcoal production close to London – close enough to cause complaint. In the sixteenth century Edmund Grindal (Archbishop of Canterbury 1575–83) summoned a collier named Grimes to appear before him. Grimes had created a nuisance by erecting a smoky kiln near the present site of Thornton Heath Station. Grindal may have been affected personally because he resided in nearby Croydon for some time and was concerned about the status of his woodlands.[14] There is little hard

FIGURE 2.2 Archbishop Edmund Grindal was concerned about his
wood supplies and the pollution from charcoal making

documentary evidence for this event, but an interesting popular tradition
seems to survive. The offence Grimes caused the Church may have allowed
him to endure in our culture. He may be the character in the title of a
sixteenth-century comedy, *Grim the Collier of Croydon: or the Devil and his
Dame, with the Devil and St. Dunstan.*[15] He appears again in a ballad of the
eighteenth century: *The Collier of Croydon.*[16] Here he proves to be a
miserable failure at seducing a young woman.

The difficulties experienced in transporting fuel were enhanced by the
poor climate of the Little Ice Age and by road conditions. The roads were
virtually unsurfaced tracks whose condition was wholly dependent on the
nature of the soil and weather. In a dry summer they would bake into a hard
surface, but in winter rain and the frequent passage of heavy wagons made
the roads all but impassable. With the growing difficulties, and consequently expense, of land carriage, transportation by water became more
and more attractive. Indeed, London's rapid growth would not have been

possible without the development of river and, more important, coastal supply routes as a vital alternative to roads. As sources of wood near navigable rivers had become totally depleted, so it was only natural that coal brought by sea from the northern ports should compete effectively. The emphasis on river transport of fuel is shown by documents relating to the Oxford-Burcot Commission set up to examine the navigability of the Thames. The preamble to a bill presented to the Lords in February 1623/4 read:

> Whereas the clearing a passage by the river of Thames to Oxford will be very convenient for the conveyance of Heddington stone and necessary to Oxford in respect to conveyance of coals and other necessaries thereto. And whereas the said passage will be very behooveful for preserving highways that are now so worn and broken, that in winter season they are for travellers dangerous.[17]

The Weald of Kent and Surrey was traditionally an important area supplying London's material needs. From the industries of this woodland region charcoal, timber, glass and iron had always been sent to London along the two main roads. These roads were also crowded with cattledrovers herding their charges along the same busy thoroughfares. The Wealden roads were often so muddy that horses were sometimes up to the girths of the saddle in mud. Driving wagons over such surfaces required strong teams of horses or bullocks. Indeed, the dire state of English roads in the late sixteenth and seventeenth centuries had considerable effect on the price of timber.

The shortage of timber in a region will naturally be reflected in increased prices. Between 1450 and 1650, while the price of timber increased, it did not increase as much as the cost of other agricultural products. Relative to inflation, timber became cheaper year by year. London, however, as well as a few other areas requiring imported fuel, did not follow this general trend; indeed, the price of timber in London increased between the mid-sixteenth and seventeenth centuries and became a particularly expensive fuel.[18] This disproportionate change was directly related to the high cost of road transportation to London and the fact that wood was in such short supply near navigable rivers in the southeast of England.

Coal as a fuel

The increasing price of timber in Tudor London forced the Londoners to use new fuels, just as it had done during the price rises of the thirteenth century. Once again coal was used to fill the ever more frequent shortages. As in the earlier crisis, the result of this change in fuel usage was a decline in the quality of London's air. In Tudor times the rise of the new fuel was

FIGURE 2.3 Briquettes of coal were thought to lower smoke production

expressed in terms of an increasing domestic use of coal. This hadn't happened in medieval times because chimneys were far less common then. Coal cannot be burnt indoors without some arrangement for getting rid of the smoke, so the use of coal in medieval England was largely industrial. Coal remained an unpopular fuel in sixteenth-century London, so during times of wood shortage it was the poor who were obliged to change over to this cheaper and dirtier source of heat. Evidence for this comes from the bequests of fuel to the poor noted in the John Stowe's *Annals*: frequently sea-coal. As the wood shortages became more severe more people used coal, but even in late Elizabethan times (when Stowe was writing) it was evident that the nobility still objected strongly to the use of the fuel. Well-bred ladies would not even enter rooms where coal had been burnt, let alone eat meat that had been roasted over a sea-coal fire,[19] and the Renaissance Englishman was not keen to accept beer tainted with the odour of coal smoke.

Such fastidiousness could not survive in the face of the inexorable economic pressure. Coal imports into London increased and its domestic use widened, and even where there appeared to be some requirement to provide charcoal as a fuel it became both too expensive and perhaps not even appreciated.[20] Complaints about the urban air pollution caused by coal through most of the Tudor period had been directed at its industrial usage. At the end of the Tudor period Queen Elizabeth found 'herself greatly grieved and annoyed with the taste and smoke of sea-coales', but this smoke arose from industrial sources.[21] An increased acceptance of coal as a domestic fuel was to come with her death, however. James VI of Scotland became James I of England. Shortages of wood and the availability of harder, less sulphurous coals from mines in Scotland had led to the fuel being used in the houses of Scottish nobles much earlier than in England, so the new king used the fuel in his household when he moved to London. No doubt this aided the adoption of coal as a domestic fuel for wealthy London households.

The effects of increasing usage of coal in London were already evident at the beginning of the seventeenth century. Hugh Platt, who wrote a book,

FIGURE 2.4 Smoky industries, such as smelting, were located in the forests of medieval England

A Fire of Coal-Balles, in 1603, said that smoke from burning coal caused damage to plants and buildings in the city of London.[22] That Platt doesn't see the problem of coal smoke as a particularly new one is in line with the fact that the use of coal as a fuel had been increasing, among the poor at least, for some time. It would seem that acceptance of the fuel by the nobility merely followed a change that was already well under way in the bulk of the population. Platt's own interest in this matter stemmed from attempts to manufacture 'coal-balles' by a process he patented in 1602. These were composed of coal and sawdust formed into a kind of briquette. Platt hoped that their use would alleviate some of London's air pollution problems. Of course it is doubtful whether they gave off less smoke than the coal alone, but schemes to manufacture briquettes for one purpose or

FIGURE 2.5 Alum-making, which rapidly adopted coal as a fuel, was
considered a polluting activity in the sevententh century

another continued throughout the first half of the seventeenth century.
None of the enterprises met with any lasting success.

A more profitable line of experimentation was being explored at the end
of the Elizabethan era. John Thornbrough, the Dean of York, was granted
a seven-year privilege 'to corecte the sulphurous nature' of coal in 1590. A
little later we find a number of Jacobean industrialists attempting to char
coal, thus removing 'the piercing and acrimonious spirits' which prevented
its use in most industrial processes apart from the 'boilings of beer or alum'.
The patent application made by Sir William Slingsby and partners in 1610
recognized the need for wood in the baking of malt, bread, bricks, tiles and
pottery and for the smelting of bell metal, copper, brass, iron, lead and

glass.[23] They realized that a successful method of charring coal might allow its industrial use to become widespread. Despite numerous experiments, coke made from coal did not become an important industrial fuel in the seventeenth century.

Coal itself found limited use in the processes of alum[24] and beer making,[25] but the result was that these industries were the subject of complaints in both Elizabethan and Jacobean times for polluting the London atmosphere. The brewing industry was a particularly frequent offender, especially in the area around Westminster.[26] The enhanced level of environmental awareness there was no doubt a result of the fact that the parliamentarians and nobles frequented that part of London.

The regulations concerning fuel use in the glass industry are particularly interesting. Traditionally glass-houses, manufacturing glass, were built in the Weald among ample stands of timber. By the end of the sixteenth century fuel shortages were beginning to have an effect on the industry. Indeed, the Court of Aldermen requested that a Venetian glassmaker, Jacob Verzelini, operating a furnace in London, should stop making glass during winter 'for sparinge of wood and fuell'.[27] Some experimentation made it possible to manufacture glass with coal by using lidded crucibles. Acts were passed under James I which obliged glassmakers to use coal as a fuel. Contemporary documents suggest that the reason for this legislation was the desire to preserve the forests of England. However, more than forest conservation was at stake. A rich courtier, Sir Robert Mansell, who had distinguished himself by raiding the Spanish coast, held the patent on the method of making glass using coal as a fuel. Despite early difficulties in Mansell's process, its use in the manufacture of glass must have earned him a tidy income and contributed to the decline of the Wealden glassmaking industry.[28]

Brickmaking also had problems in adapting to the new fuel. These seem to have been overcome by the use of dust mixed in with the coal. The dust was sometimes swept from the streets and there may be an interesting allusion to this in *A Description of the Morning*, by Swift, where he describes the early morning as a time when 'Brickdust Moll had screamed through half the street', although the *OED* would argue that the term referred to the colour of her face.

However, it is not really this industrial activity that characterizes the seventeenth-century use of coal; it is the rapidly rising domestic consumption that is most interesting. The transition from a wood-burning city to one that relied largely on imported coal had far-reaching consequences. Nef,[29] author of a classic book on the coal trade, *The Rise of the British Coal Industry*, argues for the early availability of large amounts of coal in England. It has further been argued that the high coal consumption encouraged an early industrial revolution. The coal trade had important consequences for the development of British sea-power as the

TABLE 2.2 Imports of coal into London, 1580–1680

Year	Period	Tons	Notes
1580	12 March–28 Sept.	10,785	
1585–6	Michaelmas–Michaelmas	23,867	
1591–2	Michaelmas–Michaelmas	34,757	
1605–6	Christmas–Christmas	73,984	
1614–15		91,599	One week missing
1637–8		142,579	Two weeks missing; a year of bad trade
1667–8	Midsummer–Midsummer	264,212	
1680–1	Michaelmas–Michaelmas	361,189	

Source: Nef, J.U. (1932) *The Rise of the British Coal Industry*, Routledge, London.

Newcastle–London run became an important training area for the nation's seamen. Table 2.2 shows the magnitude of the increase in coal usage in terms of coal imports over the period 1580–1680. The vessels, 'colliers', used in the coal trade underwent a major change in the first half of the seventeenth century. Initially the maximum loading was well under a hundred tons, and despite efforts to restrict growth of these vessels, their capacity increased manyfold by 1660. The increases in size did not increase difficulties in handling, because the new vessels required only ten men, less than half the crew needed to man the earlier ships.[30] This increase in capacity and efficiency was also accompanied by a corresponding increase in the overall number of ships plying the Newcastle–London route, which emphasizes the remarkably rapid growth in coal use. In 1600 there were only 400 of the smaller ships in the coal trade, while by the end of the century there were 1400 of the larger vessels carrying coal from Newcastle to London.

The famous Restoration economist Sir William Petty wrote in the middle of the seventeenth century that coal had risen to a position of complete supremacy. He also noted that previously coal 'was seldom used in Chambers . . . nor were so many bricks burned with them'.[31] By the end of the seventeenth century the supremacy of coal required no comment.

It is difficult to supply exact totals for the amounts of coal imported into London in the late sixteenth and early seventeenth centuries. However it is possible to estimate the increase as being about twentyfold between 1580 and 1680. These changes were accompanied by a significant modification in the attitudes of Londoners towards sea-coal as a fuel. A London-dweller in Henry Glapthorne's play of 1635, *The Lady Mother*,[32] is heard to complain 'would I were in my native Citty ayre agen, within the wholesome smell of seacole'. The Londoner, who had once damned sea-coal as an undesirable fuel, began to bemoan its shortage. The shortages were particularly serious

FIGURE 2.6 Chimneys became increasingly common and there was some pressure to clear the tops of the house and neighbouring buildings

during Charles I's attempts to sell the monopoly on the sea-coal trade and during the civil war the shortages were such that people suffered bitterly for want of 'sweet sea-coal'.[33]

The domestic acceptance of the fossil fuel is also reflected in the increase in the number of chimneys in the city. Harrison,[34] one of the contributors to *Holinshed's Chronicles*, which Shakespeare drew upon so heavily for his plays, wrote as a marginal note that the number of chimneys had increased greatly since his youth (mid-sixteenth century). In those times, he wrote, smoke indoors had been regarded as hardening the timbers of the house and as a disinfectant to ward off disease.

Coal played an integral role in the rapid growth of late sixteenth- and early seventeenth-century London. As C. H. Wilson has stated in his book

England's Apprenticeship 1603–1763 (1965), coal 'was an enabling condition [for] the growth of London . . . without coal the citizens would neither have kept warm nor fed themselves, nor have been supplied with the necessities and luxuries that made city life tolerable, let alone preferable'.

Notes

1. Atkinson, B. W. and Smithson, P. A. (1976) 'Precipitation', in Chandler, T. J. and Gregory, S. (eds), *The Climate of the British Isles*, Longman, London.
2. A map of rainfall variability is given in Bilham, E. G. (1938) *The Climate of the British Isles*, Macmillan, London, and Gregory, S. (1955) 'Some aspects of variability of annual rainfall over the British Isles for the standard period 1901–1930', *Quart. J. Roy. Met. Soc.*, 81, 257–62.
3. Shellard, H. C. (1976) 'Wind', in Chandler, T. J. and Gregory, S. (eds), *The Climate of the British Isles*, Longman, London.
4. Lamb, H. H. (1972) *Climate: Present, Past and Future*, Methuen, London.
5. Barton, N. (1962) *Lost Rivers of London*, Phoenix, London.
6. Swift, J. (1711) *Miscellanies in Prose and Verse*, John Morphew, London.
7. Oke, T. R. (1978) *Boundary Layer Climates*, Methuen, London.
8. Gill, D. and Bonnett, P. (1973) *Nature in the Urban Landscape*, York Press, Baltimore.
9. Manley, G. (1952) *Climate and the British Scene*, Collins, London; Evans, J. G. (1975) *The Environment and Early Man in the British Isles*, Elek, London.
10. Rosenburg, N. J. (1966) 'Microclimate, air mixing and physiological regulation of transpiration as influenced by wind shelter in an irrigated bean field', *Agric. Meteorology*, 3, 187–224; Bridley, S. F., Talylor, R. S. and Webber, R. T. J. (1965) 'The effects of irrigation and rolling on noctural air temperature in vineyards', *Agric. Meteorology*, 2, 373–83.
11. Nichols, J. G. (ed.) (1852) *Chronicle of the Grey Friars of London*, Camden Society, vol. 53.
12. The best-known account of this comes from the monumental Nef, J. U. (1932) *The Rise of the British Coal Industry*, Routledge, London; Flinn, M. W. (1959) 'Timber and the advance of technology: a reconsideration', *Annals of Science*, 15, 109–20, suggests that the timber famine may have been exaggerated.
13. O'Riordan, T. and Turner, R. K. (1984) 'Pollution control and economic recession', *Marine Pollution Bulletin*, 15, 5–11.
14. Fitter, R. S. W. (1945) *London's Natural History*, Collins, London. I have not been able to trace a primary reference to Grindal's meeting with Grimes. There is plenty of evidence about Grindal's life in Croydon and he died there in 1583. He was much out of favour with Elizabeth I and this led to concern over the wood from his forests: Garrow, D. W. (1818) *The History and Antiquities of Croydon*, Croydon and Strype, J. (1821) *The History of the Life and Acts of Edmund Grindal*, Clarendon Press, Oxford.
15. Tatham(?), J. (1662) *Grim the Collier of Croydon: or the Devil and his Dame, with the Devil and St. Dunstan*, London. There is evidence that this printed version of

the play is very much later than the original. The office book of the Master of Revels records that a play called *The Historie of the Colyer* was performed by the Earl of Leicester's men in 1576. It appears that Grim may be identified with a character of the same name in an earlier play, *Damon and Pithias* (1571). These earlier dates are contemporary with Grindal, although he was not made Archbishop of Canterbury until 1575.

16. In Holloway, J. and Black, J. (eds) (1975) *Later English Broadside Ballads*, Routledge & Kegan Paul, London. Colliers were none too popular because of the way they manipulated the prices of coal (see Platt or Nef references, note 22 and note 12), so they became the subjects of ballads, e.g. H. D. and T. C. (1720s) *The Battle of the Colliers*, London Brit. Lib. Cat. c.116.i.4(23).

17. Thacker, F. S. (1968) *The Thames Highway*, vol. I, reprinted David & Charles, Newton Abbot.

18. Wilson, C. H. (1965) *England's Apprenticeship 1603–1763*, Longman, London. This is an excellent book which includes discussions of many of the topics treated in the latter part of this chapter: the problems of bad road conditions and those of industry in adapting to coal as a fuel, coastal trade, wood prices, increasing coal usage.

19. Stowe, J. (1592) *Annals of England*, R. Newbery, London; Howes, E. (1631) *Annals of England Continued by E. Howes*, printed by A. M. for R. Meighen, London.

20. See *The Endowed Charities of the City of London*, Sherwood, London (1829). The will of John Costen (Costyn) of 1442 required that the poor of the parish of All Hallows Staining be given 100 quarters of coal between All Hallows Eve and Easter Eve. From about 1633 to about 1660 the requirements of the will were strictly adhered to and charcoal was delivered. Evidently 'coal' here meant charcoal. However, after that time sea-coal was provided instead because it was 'much more suitable to their wants'.

21. *Cal. State Papers (Dom.)*, 1547–80, 612.

22. Platt, H. (1603) *A New, Cheape and Delicate Fire of Cole-balles . . .*, London and a patent of 1602.

23. For Thornbrough, see *Cal. State Papers (Dom.)* Eliz. vol ccxxiii (10 Oct. 1590); for Slingsby: Landsdowne MS 67 no. 20.

24. *Cal. State Papers (Dom.)*, 1603–61, 625, 1611–19, 13; see also Nef (note 12 above).

25. *Cal. State Papers (Dom.)* Charles I 1627–8, 269–70; Brimblecombe, P. (1976) 'Attitudes and responses to air pollution in medieval England', *J. Air Poll. Control Assoc.*, 26, 941–5.

26. See, for example, *House of Lords Calendar* (16 Feb. 1640–1).

27. Sutton, A. F. and Sewell, J. R. (1980) 'Jacob Verzelini and the City of London', *Glass Technology*, 21, 190–2.

28. Godfrey, E. S. (1975) *The Development of English Glassmaking 1560–1640*, Clarendon Press, Oxford; Kenyon, G. H. (1967) *The Glass Industry of the Weald*, Leicester University Press.

29. Nef, J. U. (1932) *The Rise of the British Coal Industry*, Routledge, London.

30. ibid., vol. I, 391. Smaller crews meant higher profits, but also meant that vessels were useful for long voyages; Cook's *Endeavour*, for instance, started life as a collier. The rapid rise in size of vessels caused some concern and there

were attempts to limit the size of colliers (see *Cal. State Papers (Dom.)*, 1625–26, 311).

31. Hull, C. H. (ed.) (1963–4) *The Economic Writings of Sir W. Petty*, Kelly, New York.
32. Glapthorne, H. (1635) *The Lady Mother*, Malone Society Reprint (1958).
33. 'A recipe for making briquettes' (1644) British Museum, 669, f. 10 (11).
34. Harrison, W., (1577) in *Holinshed's Chronicles*, Book III, c. 10.

3

Evelyn and his circle

Writers of the mid-seventeenth century such as John Evelyn were critical of the increasing use of coal as a domestic fuel in London. Their criticism came after two centuries in which there seem to have been relatively few complaints. As we saw in the last chapter, James I of England appears to have been indirectly responsible for increasing the popularity of the fossil fuel. Legislation, grants of patents and the use of coal in the royal household all served to promote the fuel. However, the King was also concerned about the effect that coal smoke had on the buildings of London. In 1620 he was 'moved with compassion for the decayed fabric of St Paul's Cathedral near approaching ruin by the corroding quality of coal smoke to which it had long been subjected'.[1] The King's concern about the evils of smoke also appears in his *Counterblast to Tobacco*,[2] which not only expressed disgust at the increase in smoking, but also touched on the matter of pollution when he condemned the besooted state of English kitchens.

The comments about pollution in the seventeenth century came with a period of rapid growth in science, particularly experimental science. Scientists were strongly influenced by the ideas of Francis Bacon, who insisted on the importance of careful experimentation and deduction. It is a characteristic of the earliest works on air pollution, which began to appear in this period, that their authors take remarkable care in observation. The philosophical developments in science during the early seventeenth century were accompanied by improvements in scientific communication, notably with the development of the scholarly journal. Hitherto scientists often preferred to guard their discoveries in the hope of material gain, but the gentlemen scientists of the seventeenth century found intellectual challenge and widespread fame more rewarding than financial profit. Learned bodies, such as the Royal Society, which received its charter in 1660, also fostered scientific advances.

In terms of the scientific study of air pollution, the most notable figure of the seventeenth century was John Evelyn. Although today his fame is more as a literary figure and centres largely on his diaries, which describe many aspects of London life between 1641 and 1706, his other writings have been receiving increasing attention from environmentalists today, who have begun to regard him as something akin to the field's patron saint. His portrait has adorned the cover of *Clean Air*, the journal of the National Smoke Abatement Society, and has appeared on their headed notepaper. Especially interesting to environmentalists are his contributions to urban planning and architecture, to silviculture and to air pollution science.

He was an active member of the Royal Society and served as its secretary for a short time. His book on air pollution, *Fumifugium*,[3] is an outstanding work and cannot fail to remind the reader that he was a man with extraordinary powers of perception. However it must not be considered in isolation because it reflects much that was being written and said by influential people of the time. After all it was an age that actively encouraged scientific discourse.

In addition to being personally aquainted with contemporaries interested in air pollution, Evelyn was aware of both the classical sources of knowledge in the field and the earliest English writers on the subject. In his book *Sylva*[4] he refers to Hugh Platt's *A Fire of Cole-Balles* (1603)[5] which we have already mentioned briefly. One of Platt's chief concerns was the provision of useful labour for the unemployed of London, a familiar theme for the period. Platt hoped that the poor might gain income by kneading coal-balls from a thin paste of loam and poor grade coal. In addition it seemed possible that the fires made from such fuel would not be as offensive in smell, or as soiling, as those made from ordinary sea-coal. He was also concerned with the damage that the smoke caused to the gardens, furnishings and clothes of the nobles of London, claiming that the problem had been widespread even before the turn of the century. It reminds us that the use of coal was not without consequence in Elizabethan times.

Apart from Platt, Archbishop Laud (1573–1645) was probably the early seventeenth-century figure who most influenced John Evelyn's interest in air pollution. Although this influence is rather indirect, Laud proves to be one of the most interesting characters concerned with air pollution in late Jacobean London. However, William Laud may not have been the first archbishop to have been worried about the pollution of the air. Much early action against air pollution came from the Church and some modern authors have claimed that the frequent protests against the use of coal may all be part of an association between sulphurous fuels and anti-clerical forces.[6] However it is likely that this reaction simply reflects the broad powers of the Church at this time. After all, we saw in the previous chapter the evidence that Archbishop Edmund Grindal complained about the smoke nuisance from charcoal production. If we have indeed correctly identified the Collier of Croydon as a charcoal maker, then the emissions would hardly have been sulphurous.[7]

We have good evidence of Laud's interest in air pollution because he was impeached and sent to the Tower of London in 1640 and some of his

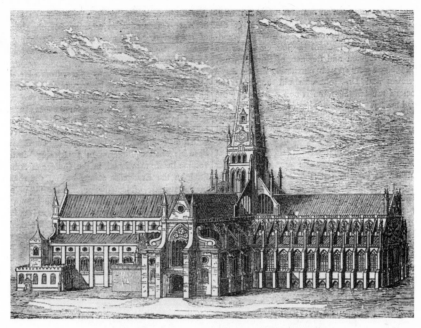

FIGURE 3.1 Old St Paul's was severely damaged by the corroding quality of coal smoke

activities in the field of smoke control appear in the accounts of his trial for treason.[8] The Archbishop had made himself extremely unpopular by following a strict line in both religious and legal matters. It appears that he had made the brewers of Westminster pay fines for using sea-coal, even though the King had pardoned them. Part of the money Laud had extracted from the brewers was to be used to repair St Paul's Cathedral, which was suffering badly from the polluted state of London's air. Although one cannot find much to admire in Laud's administration, it did pay overwhelming attention to detail; the damaging effect of the easterly drift of coal smoke from Westminster towards central London can hardly have escaped its notice. The desire to make the polluter pay for the damage this smoke caused would receive a more sympathetic hearing today than it did in the seventeenth century.

Despite the fact that these activities were used as evidence against the Archbishop, he was not alone in exerting pressure on the brewers of Westminster to return to a less objectionable fuel than coal. In the opening half of the seventeenth century these brewers were likely to have come into conflict with a confusing range of offices, including those of the Attorney-General, the judges of Serjeant's Inn and Inner Temple and the Court of Exchequer, in addition to the ecclesiastical bodies.[9] The reason for the

FIGURE 3.2 Archbishop Laud, attentive but authoritarian, fined brewers in Westminster for emitting coal smoke

confusion may have arisen from semi-legislative agreements that had developed over the previous hundred years. The Company of Brewers had understood, in 1578, that Her Majesty was 'greatly greved and annoyed with the taste and smoke of the sea-coales' they used and offered to burn only wood in the brewing houses near the Palace of Westminster.[10] This agreement reached in Elizabeth's reign was obviously a voluntary one, so it is not surprising to find attempts to have it formally converted into a law.[11] It was brought before Parliament as a bill in the 1620s; the act was passed by the House of Lords in 1623, forbidding the use of sea-coal in brewhouses within one mile of any house within which His Majesty's Court or the Court of the Prince of Wales was habitually held. Despite its passage through the Lords, the Commons dropped it at the end of the session. Thus the fines and bonds Laud levied against brewers in the 1630s and 1640s, which often amounted to several hundred pounds, were of doubtful legality. William Laud was found not guilty of treason by the House of Lords, but he was convicted under a bill of attainder by the Commons in 1644 and executed the following year.

Sir Kenelm Digby, FRS (1603–65)

In the late 1630s, while Charles I tried to sell the monopoly on the coal trade, and later during the civil war, London was frequently short of coal. This allowed the inhabitants to appreciate the great difference that extensive burning of fossil fuel made to the air of the city. Among the Royalists who fled the city with the coming of Parliamentary power in the 1640s were John Evelyn and Sir Kenelm Digby. Both men were destined to join the young Royal Society and to make important observations about the pollution of London's air. Their sojourn on the Continent and more particularly the time they spent in Paris, which burnt very little coal, must have been a crucial experience in the growth of their attitudes towards air pollution. They met several times during this period of exile, and Evelyn visited Digby's laboratory in Paris where he was engaged in alchemical experiments, attempting to dissolve gold in 'exceedingly rectified rain-water of the autumn equinox'.[12] They moved in the same circles in later years. There seems little doubt that air pollution would have been one of the many topics that they discussed together. After one of their early meetings on the Continent, Evelyn added a note to his diary that Digby was an 'arrant Mountebank' (i.e. a downright charlatan). While this shows something less than complete admiration for the man's work, it probably refers to his speculative alchemical notions rather than his observations on air pollution, which Evelyn was to use later.

It is particularly fascinating to note that Archbishop Laud was Sir Kenelm Digby's tutor before he fled to the continent. It is tempting to imagine that Digby and Laud may have talked at some time of the problems that arose from the coal smoke in London's air. Digby left England in 1643 and was not to return for more than a decade; his friendship with Laud, the fact that he was a Catholic and his father's involvement in the Gunpowder Plot all made this essential. The long absence from London would have meant that his ideas on the polluted state of the city's air must have been much influenced by recollection of past experiences. His vision of London's air pollution would have been contemporary with that of Laud. During Sir Kenelm's brief visit to England in 1654 he stayed with John Evelyn. Before his final return to England, a few years later, he had a number of pamphlets published to emphasize his patriotism. Among these was *A Late Discourse made to a Solemne Assembly of Nobles and Learned Men at Montpellier in France, by Sir Kenelme Digby, Knight &c. Touching the Cure of Wounds by the Powder of Sympathy; with instructions how to make the said Powder; Whereby many other Secrets of Nature are unfolded.*[13] The title of this book is sometimes abbreviated to *A Discourse on Sympathetic Powder*.

A Discourse on Sympathetic Powder, one of his best-known works, is a tiny book. It ranges from discussing 'an admirable history of a Nunne at Rome that was troubled with wind' to a 'discourse on gold and quicksilver'.

FIGURE 3.3 Sir Kenelm Digby, 'arrant mountebank' and speculative
scientist, believed that London's coal smoke caused deaths from 'ptisicall
and pulmonicall distempers'

Despite such idiosyncrasies it is rather modern in one way. The
observations Digby makes are used to justify an atomic theory of matter.
He tries to convince the reader, through very commonplace examples, that
matter is very finely dispersed as minute corpuscles or atoms. The far-
ranging nature of his discussions did not help clarity. John Beale, FRS
(remembered for his *Aphorisms concerning Cider* that are to be found
appended to Evelyn's *Sylva*) complained of Corpuscularian Philosophy that
he had long given it up, as Sir Kenelm Digby had made him quite giddy
with it.[14] At the beginning of the seventeenth century Hugh Platt had made
some attempt at a semi-scientific explanation of the problems caused by
sea-coal smoke. Digby tried to apply an atomic theory to air pollution and
postulated that the damaging effects of coal smoke arose from the fact that
the atoms were sharp and pointed. Atomic theories were in vogue in the
mid-seventeenth century but there do seem to be some interesting roots to
Digby's dizzy atomism.

While in exile in Paris Digby moved among the English Royalists
staying there. Among them was Margaret Lucas who married William
Cavendish (later the Duke of Newcastle) in 1645. The wedding service
took place in the private chapel of Sir Richard Brown, John Evelyn's

FIGURE 3.4 Margaret Cavendish, eccentric in both fashion and thought, proposed an atomic theory of coal combustion

father-in-law who was then English Resident at the Court of France. Sir Kenelm Digby may not have been present at the wedding, but it is known that he gave the Duke of Northumberland a perspective glass. Margaret eventually became a prolific although somewhat eccentric writer of scientific theories. She had a temperament akin to Digby's[15] and her speculative scientific reasoning ran parallel to his.

In 1653 she published her first book, *Poems and Fancies*.[16] It was probably never widely read; the poetry is awful and its wilder speculations are well out of line with the rigour of seventeenth-century scientific thought. Nevertheless she was in tune with miasmatic theories for the origin of diseases and wrote that 'when the air is corrupted, it produces several diseases'. This was by no means an original notion. More significant is the fact that she proposed an atomic theory for the combustion of coal:

> Why that a *Coale* should set a house on *Fire*
> Is, *Atomes sharpe* are in that *Coale* entire,
> Being strong armed with *Points*, do quite pierce through;
> Those *flat dull Atomes*, and their *Formes* undo.

These ideas may have contributed much to Digby's thinking about the time that he wrote *A Discourse on Sympathetic Powder*. Certainly in 1657, the year before the discourse was published, he was reading Margaret Cavendish's scientific writings. He wrote a letter to her then,[17] thanking her for a copy of a book which might have been either *Poems and Fancies* or *Philosophical and Physical Opinion* (1655). Her ideas seem to contribute to Digby's belief that the smoke from coal contains a very volatile salt, whose sharp atoms filled the London air. If we accept an idea then current that considered acids to have sharp needle-like atoms and alkalis to be hollows, such that neutralization occurred by blocking off the sharp points, then we may admire Digby and Cavendish's approach even more. We now know that the corrosive properties of atmospheres polluted with coal smoke come largely from the production of sulphuric acid; so attributing the damaging effects of air pollution to sharp atoms, or rather acids, was not far wrong.

Digby was particularly worried about the damage these volatile and corrosive salts caused to the lungs, claiming that their presence in the London air was responsible for the high incidence of deaths from pulmonary complaints in the city. It was argued that more than half those who died in the metropolis did so from 'ptisicall and pulmonary distempers, spitting blood from their ulcerated lungs'. The air of London was worse than that of Paris or Liège, so people with weak lungs and plenty of money were advised to live on the continent. This argument was in contrast to the one offered by Sir William Petty, FRS (1623–87) who thought the air of London was more wholesome, and that there were fewer deaths than in Paris, because the fuel of London was cheaper and not as bulky, being a wholesome sulphurous bitumen.[18] No doubt he thought that a fuel that was cheap and compact would be more readily purchased and lead to warm, safe interiors in the winter. Digby would have disagreed, although he was willing to acknowledge that Paris suffered excessively from stinking dirt which mingled with the air. He insisted that the Parisian air was not as pernicious as that of London. Most of the air pollution in Paris arose from the rotting wastes in the sewers of the city but, as previously noted, such olfactory pollutants usually received far less attention than those from coal burning. Digby's main concern in *A Discourse on Sympathetic Powder* is with human health; his observations on other effects of air pollution are limited to the fairly obvious bespeckling of tapestries, valuables, clothes and plants.

John Evelyn, FRS (1620–1706)

Digby's fellow Royalist, John Evelyn, wrote the classic *Fumifugium* or *The Inconveniencie of the Aer and the Smoak of London Dissipated* in 1661. It acknowledges Sir Kenelm Digby's book as a source for some material and shows a clear development from some of Evelyn's own writings, such as *Sylva* and the satirical tract *A Character of England*.[19] In the latter work he advanced ideas on the need for extensive urban planning to make cities attractive places for their citizens. He was influenced by his experiences on the continent gained during the years of exile. His early works are tinged with youthful enthusiasm for change: *A Parallel of Architecture* (1664) shows an interest in the city as an attractive place to live. Evelyn found the streets of London composed of a congestion of misshapen and extravagant houses in streets much too narrow. He claimed that public works lagged far behind those found in continental cities. In England leaders were more likely to strike a medal or build a monument than to repair an old bridge or widen a street.

Writing some time after the publication of *Fumifugium*, Evelyn terms the work an *invective* (*Diary*, 4 September 1666), but this should not be taken to mean that it is an aggressive, ill-considered piece of writing. Its author suggests in his introduction that he was driven to write the tract when indignant at the difficulties he experienced trying to view the palace of Charles II at Whitehall, surrounded and infested, as it was, by smoke from chimneys near Scotland Yard. While the reasons Evelyn gives for writing *Fumifugium* may be true, in part, much is mere protocol, as Evelyn desired to obtain the King's backing for his far-sighted plans. The real impetus behind the work came from long observations of the detrimental effects of the smoke-laden air. He regarded coal as a 'sullen' fuel and preferred wood or charcoal. In 1656 on his return from East Anglia he had visited Sir Joseph Winter who was experimenting with the charring of coal (*Diary*, 11 July 1656). He had seen the coal heated in earthen crucibles until it could be withdrawn as a mass of half exhausted cinders. When rekindled this charred coal made a clear and pleasant chamber fire. Evelyn did not forget this in his later writings, although he must have regretted the consistent lack of enthusiasm that coke met with.

It is probably in *A Character of England* that one finds the keenest display of his early interest in air pollution. He writes that the city was cloaked in

such a cloud of sea-coal, as if there be a resemblence of hell upon earth, it is in this volcano in a foggy day: this pestilent smoak, which corrodes the

FIGURE 3.5 John Evelyn, 'old Sylva', was England's first environmental
radical

very yron, and spoils all the moveables, leaving a soot on all things that it
lights: and so fatally seizing on the lungs of the inhabitants, that cough
and consumption spare no man.

Evelyn could see no excuse for allowing London's air to be so bad. The
city had been built on 'a sweet and agreeable eminency of the ground', with
a gently sloping aspect that allowed the sun to clear the fumes from the
waters or lower grounds lying to the south. He assumes that the 'Antient
Founders' of the city carefully considered the need for ventilation when
choosing an urban site. All this, he thought, took place long before Julius
Caesar arrived in the British Isles or Caesar's engineer Vetruvius
expounded these very principles in his books on architecture. Today,
although we must reject the mythological foundations of London, it is
hardly possible to consider the city badly sited for clean air. It has none of

the problems of the Los Angeles basin where high mountains and sea breezes trap the air under persistent inversions. Despite the advantageous prospect, John Evelyn found London covered with a 'Hellish and dismall Cloud of SEA-COAL'. The blame for this is laid squarely on the owners of some 'few funnels and Issues, belonging to only *Brewers, Diers, Lime-burners, Salt* and *Sope-boylers*, and some other private trades, *One* of whose *Spiracles* alone, does manifestly infect the *Aer*, more, than all the chimnies of London put together besides'.

The effects of smoke, as described in *Fumifugium*, were numerous: churches and palaces looked old, clothes and furnishings were fouled, paintings were yellowed, the rains, dews and waters were corrupted, plants and bees were killed, and human health and well-being were ruined. Evelyn did not simply draw attention to the increased death rate noted by Digby, but emphasized the general decline in health that was brought about by a smoky atmosphere. While admitting that in mines and foundries the workers were exposed to extremely smoky atmospheres yet survived, he says that we could hardly commend this type of existence. Here Evelyn seems to touch on a principle still very much in evidence in our modern legislation. It is often accepted that in some occupations there is a risk and that somewhat higher levels of exposure to toxins of radioactive material may have to be tolerated. However, levels encountered in occupational exposure are not necessarily acceptable to the population as a whole.

Evelyn realized that the solution to the problems of the metropolis was not simple. He reminds his readers of the phrases 'no smoke without fire' and 'no fire without smoke'. By challenging the second of these Evelyn was to put forward the essence of a smoke abatement policy that took two centuries to be adopted. While smoke might not be found without fire, it was just possible that one might have fire without smoke, or at least without very much smoke. He thought that one way of achieving this would be for the houses of London to be supplied with wood, which is a much more pleasant domestic fuel. Then there were the possibilities of charring either wood or coal. However it was evident that these measures would not be enough because Evelyn, who felt very strongly that it was industry that caused the main problem, proposed that an act be laid before Parliament to remove the obnoxious industries to the outskirts of London. These industries would be 'removed five or six Miles, or at least so far as to stand behind that Promintory jetting out, and securing *Greenwich (or Wooledge)* from the *Aer* or *Plumstead-Marshes*'. It would seem that he is using the word 'promintory' to mean the bend in the Thames or a hill behind which industries could nestle. The site envisaged probably lay beneath Shooter's Hill (Fig. 3.6). The increased traffic on the River Thames was to provide employment for London's numerous watermen. Where industry had to remain closer to London it could be situated in 'The Town of Bowe' where continual winds provide a partial cure to the problem.

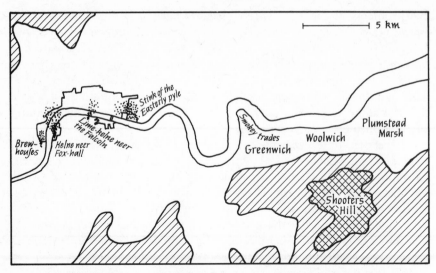

FIGURE 3.6 Map of Restoration London showing sources of air
pollution noted by John Evelyn and suggested location for smoky trades

The final section of *Fumifugium* contains a discussion of the malodorous
emissions from other trades, such as chandlers and butchers, where large
amounts of decomposing rubbish frequently gave offence. These trades
were also to be removed beyond the city walls. Finally, as a positive step
towards the purification of the air, odiferous flowers, plants and hedges
were to be planted about the city.

On 14 September 1661, *Fumifugium* was presented to King Charles II
who was apparently happy that it be 'Published by His Majesties
Command'. However, these words were deleted from the title page of the
second issue which came from the presses in the same year. Perhaps Charles
II was already having a change of heart. Evelyn discussed the subject
broached by *Fumifugium* with his Majesty on 1 October and was
commanded to prepare a bill to place before Parliament. The draft of this
bill was drawn up by the Queen's Attorney, Sir Peter Ball, and was in
Evelyn's hands by 11 January 1662, according to his *Diary*. The bill would
have removed smoky trades from the city, but nothing more is heard of it
so we must assume that it was dropped. Evelyn was no doubt disappointed
with this and the lack of perseverance in environmental matters displayed
by England's monarch. It was probably financial considerations that
defeated Evelyn's schemes, as nowhere in his writings does he show
awareness of the economic problems inherent in his plans for reconstruction.[20] Environmental idealism often runs up against economic
considerations.

Evelyn did not give up. In the following years he took further interest in
the architecture of London and served as a commissioner for improving

FIGURE 3.7 Digby complained that soot fouled washing set out to dry

city streets from May 1662. However, he seems to have been a rather inactive commissioner and attended only three meetings during his term of office.[21] Nevertheless at least one part of his plans outlined in *Fumifugium* met with some success: this was the suggestion for the construction of parks planted with sweet-smelling flowers.

The best opportunity for reconstruction came after the Great Fire destroyed so much of the city in 1666. Evelyn claimed that many people saw his visions in *Fumifugium* as prophetic: he had warned of the dangers of fire inherent in the industries of London.[22] In the weeks following the fire he mapped the areas of the city that had been destroyed and presented an outline for their rebuilding to Charles II. His influence can also be noted in the *State Papers*[23] immediately after the fire, where it is suggested that in reconstructed London 'brewers, bakers, sugar bakers and others whose trades are carried on by smoke are to live together in the same quarter assigned to them'. This certainly does not go as far as Evelyn would have liked; complete removal of such trades from the boundaries of the city would have been the solution he most desired. Despite his influence and that of a number of far-sighted men of the time, the coherent plan for a new London was not followed. The reconstruction that took place was haphazard and London's air pollution continued unabated.

John Graunt, FRS

Both Evelyn and Digby were convinced that the polluted atmosphere of London took its toll of the lives of its citizens. Digby did not give any source for his information on London's death rate. However Evelyn, in a section of *Fumifugium* discussing the work of *Sir K.* (Digby), says the evidence may easily be seen in the Bills of Mortality. The Bills of Mortality were an early attempt at keeping medical statistics, encouraged by the incessant plagues that had ravaged London through previous centuries. The earliest of these bills date from 1532, and at first they were only published during times of plague. However, during the reign of Elizabeth I they became a regular series, and so contain a wealth of medical information about the health of Londoners.

While the two Fellows of the Royal Society, Digby and Evelyn, were writing their books and showing concern for the state of London's atmosphere, John Graunt, a draper, was gathering together all the statistical information from the London Bills of Mortality for analysis. His work was completed and published as *Natural and Political Observations . . . Made upon the Bills of Mortality* in 1662.[24] This book was to have a great impact. Its merit was so evident that Charles II recommended Graunt's election to the Royal Society, adding the comment that 'if they found any more such tradesmen, they should be sure to admit them all without further ado'. This may well reflect the King's feeling that the Royal Society was promoting too many unprofitable experiments such as the 'weighing of air' by Robert Boyle in attempts to understand the nature of gases.[25]

Graunt's book remains a pioneering work of demography. No mere catalogue of the number of deaths within the city, it shows an awareness of the inherent statistical problems in taking an enormous data set and trying to make general statements about it. Graunt realized that the amount of accumulated data was so large that it was impossible for the reader to appreciate it. It was necessary, somehow, to reduce it to a simpler form. This involved re-tabulation and simple calculations, which allowed him to draw attention to the important aspects of the figures. He also looked at the underlying nature of the data itself, asking such questions as: Was it reliable? How was it collected?

FIGURE 3.8 The searchers, old women who ascertained cause of death

The Bills of Mortality were published weekly from figures returned by the parish clerks, who had assembled their data from reports made by *searchers*. These were 'ancient Matrons', sworn to the office, who examined all the bodies within the parish to determine the cause of death. It is obvious that their medical background would have been both rudimentary and variable, so it is understandable that the causes of death given in the reports were frequently vague and inaccurate. Graunt was aware of the effect that inaccurate observations might have on his work. For instance, the rise in the number of deaths from rickets was of particular interest to his studies (Fig. 3.9). It was not possible to know whether the increase was due to the fact that the disease was a relatively new one or whether it merely reflected greater experience in diagnosis. The physician Francis Glisson[26] was of the opinion that it was a new disease. Glisson was the author of *Tractatus de Rachitide*, a treatise on the disease which is considered to be one of the finest pieces of medical writing of the century. He probably knew more about rickets than anyone else in England at the time. The fact that he thought the disease was a new one, although it actually occurs sporadically in even the earliest human populations, illustrates the striking increase in its prevalence that must have characterized the first half of the seventeenth century. The notion of new diseases arising suddenly, achieving prominence and then vanishing was much in vogue at the time. This had happened in previous centuries with diseases such as English Sweat. Such conclusions are understandable when the availability of long-term medical records was limited to classical sources that referred mostly to warmer climates.

Over many historical periods rickets was quite rare or restricted to areas of extreme poverty. It is essentially a nutritional disease arising from a lack

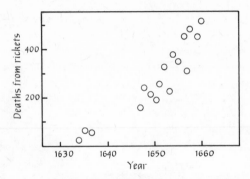

FIGURE 3.9 The incidence of deaths from rickets in seventeenth-century
London

of vitamin D. While the fault is usually dietary, it may also be due to lack of
exposure to sunlight. The calcium metabolism is upset, and bones,
particularly among the young, can soften. One cannot easily assess the
various factors that controlled the incidence of rickets in Stuart London,
where even the daughter of Charles I, Princess Elizabeth, may have been
affected,[27] but it is interesting to note that the rise of the disease parallels the
increasing use of coal in London. Presumably a number of interlinked
factors could have helped to increase the frequency of the disease: (i) the
almost complete covering of the skin required by the cold weather of the
Little Ice Age; (ii) the poor quality of the food brought about by shortages
and the attendant rise in prices; and (iii) the foggy soot-laden skies. The
grey skies of the time, which must have cut off a large portion of the winter
sunlight, received ample comment from travellers. H. R. Bentham, who
visited London in the second half of the seventeenth century, wrote that the
English have 'thick skies and cloudy weather'.[28] It would seem that Graunt
had, by accident, noticed the increase in the prevalence of a disease which
can be induced by a polluted urban atmosphere. However, contemporary
medical science could not have provided him with the knowledge required
to link rickets with the obscuration of ultraviolet radiation by the
particulate materials suspended in London's air.

The tragedy is that the cause of rickets and the simplicity of its
prevention and treatment, through adequate diet, were not realized until
late in the nineteenth century. At this time more than half the children
dwelling in English cities suffered from the disease. The smoke-filled skies
of Victorian cities can no doubt be implicated in the remarkable incidence
of rickets[29] but today the frequency of bone diseases among the young is
very much lower. It would be wrong to attribute these changes to cleaner
air, however; improved diet has made the largest contribution to lowering
the frequency of rickets. If we look at bone disease among older people
there is some statistical evidence to suggest that air pollution may be linked

with a weakening of their bones. A correlation between the levels of smoke in the air and deaths from fractures after falls has been shown by Eddy.[30]

In recent years there has been some concern about the number of cases of rickets among immigrant children. It was assumed, at first, that this was due to the pigmentation in their skin, which made them less able to utilize the scant sunshine available in the British Isles in winter. However, more recent evidence points once again to dietary causes. Diets with large amounts of carbohydrate cause bone growth to be rapid. In some cases this may be too rapid for the calcification processes to take place properly. Another cause may be related to the type of carbohydrate. The diet of Asian immigrant children is often dominated by chapatis, which are made from an unrefined carbohydrate. It has been suggested that this tends to complex calcium in the gut and prevents its absorption into the blood stream where it can ultimately be used in bone formation.[31]

It is difficult to link specific diseases, such as rickets, to air pollution. Graunt attempted to make more general comments about the effect of sea-coal on human health. It was evident from the Bills of Mortality that after the restoration of Charles II the death rate in London was much higher than in rural areas. Looking at the first half of the century as a whole he gave the following figures: between 1603 and 1644 there were 363,935 burials in London, while only 330,747 christenings; in a country parish, probably in Somerset, there were 5280 burials compared with 6339 christenings. Graunt felt that there were a number of reasons for the death rate in London being higher than the birth rate. He thought that the birth rate would be lowered by a number of factors: (i) those in London for business or pleasure frequently left their wives in the country: (ii) apprentices were not able to marry; (iii) seamen's wives dwelt in the city; (iv) there was much adultery and fornication in the city; (v) business in London was full of stress.

However, he also considered the death rate in London to be higher than in the country. The stinks and airs of London were thought to be an important contributor to mortality. The unhealthy character of the air was largely blamed on the presence of sea-coal smoke which had increased so much throughout the seventeenth century. The smoke was not simply unpleasant: it caused 'suffocations which many could not endure', making London's death rate so much worse than that in the country.

These conclusions are interesting, although it is hardly likely that they are fully correct. To blame the high death rate entirely on the pollution of the atmosphere was to ignore numerous other environmental factors: the widespread malnutrition, unhygenic conditions, poor post-natal care and a high population density, which allowed a rapid spread of diseases that could not take hold in more dispersed rural populations. A more important aspect of Graunt's conclusions is the light they throw on the general attitudes of the times. Graunt was a draper, which meant that he would

TABLE 3.1 Figures presented by John Graunt to show the rapid fluctuation of the death rate due to fluctuations in the airs.

Week	Deaths
1	118
2	927
3	993
4	258
5	852

Source: Graunt, J., *Natural and Political Observations . . . Made upon the Bills of Mortality*, London (1662).

have lacked a training in medicine that would have enabled him to keep abreast of recent medical discoveries. His ideas probably came from beliefs that were fairly widely held during the mid-seventeenth century. Graunt's *Natural and Political Observations* reminds us that the people of the seventeenth century believed that diseases arose from the spread of contagions through the air. As already mentioned, theories of this type had been prevalent since classical times and Graunt thought that he had found evidence to support the theory of the miasmatic origins of the plague in the rapid fluctuations in death rate that occurred over weekly intervals (Table 3.1). He believed that these changes could be related to changes in 'airs', but he does not attempt to support his thesis with any meteorological data as his study was really complete before the Royal Society and a number of contemporary English physicians began to take an interest in recording meteorological data.[32]

In 1664 Nathaniel Henshaw, also a member of the Royal Society, published *Aero-chalinos*[33] in Dublin; a curious work concerned with the need for fresh air for a healthy constitution. It shows the interest in the relationship between the atmosphere and health that was becoming widespread and also explains why the early development of meteorology was so closely linked with medicine.

However, the interest in meteorology was taken beyond mere aphorisms about the relationship between weather and health. The study of the chemistry of the atmosphere, particularly that of trace components, seemed to arise out of the contemporary miasmatic approach to the origin of disease. As there was no germ theory at the time it was natural for the scientists to postulate toxic trace elements as the contagious agents in the air. Not all trace components of the atmosphere were thought to be harmful. After all, Digby's *powder of sympathy* was an airborne material meant to have a beneficial effect. However, the relationship seventeenth-century scientists thought to exist between compounds in the air and human health made it important for them to identify such elements in the

FIGURE 3.10 There was no large-scale lead production in London, but the activities of plumbers have caused occasional complaints from medieval times to the present day

atmosphere. Robert Boyle, FRS (1627–91) suggested a number of ways in which they might be detected,[34] although it isn't likely that he made any use of these methods. When postulating atmospheric contagions of disease the early scientists suggested the materials that they already knew to be toxic, in larger quantities, as liquids or powders. Evelyn thought that the charring of coal would have important implications for human health in cities because it would remove arsenic and sulphur. A physician, John Carte, wrote to Nehemiah Grew, FRS (who became secretary to the Royal Society in 1677) about the effect on health of emissions from lead smelters. Carte thought that the disease Belland, which was indigenous to Derbyshire, arose from toxic materials present in the smoke. It was suggested that these might be trace quantities of antimony or mercury.[35]

The early attempts at determining the trace components in the atmosphere that influenced health seem quaint, but considering the difficulties still encountered in attempts to understand the effects of environmental toxins, it is easy to sympathize with these early scientists. Even today most of our knowledge of toxins is based on experiments that are run at very high pollutant concentrations rather than at those which typify the atmosphere. Knowledge of the effect of these toxins at the much lower concentrations found in the environment is often little more than informed guesswork. Thus it comes as no surprise that the early members

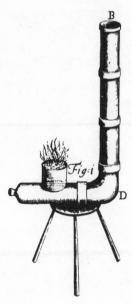

FIGURE 3.11 Justel's smokeless stove which could consume sardine oil
or cat's urine without offence

of the Royal Society chose known poisons such as lead, arsenic and
antimony as the malignant agents in polluted air.

Although the Fellows of the Royal Society debated about the origin of
disease and the quality of London's air, they seem to have made little
practical contribution to abating smoke. One exception may have been
Henri Justel, FRS, who described a stove which would consume its own
smoke[36] and published an illustration in the Society's journal (Fig. 3.11). In
this device, once the chimney of the stove was very hot, the smoke would
pass down through the coals, which meant that it was burnt in much the
same way as in more sophisticated stoves and firing systems developed in
the nineteenth century. This particular stove was said to be so successful
that 'coal steept in Cats-piss makes not the least ill scent'.

Historical evidence

Much of the statistical material available to Graunt still exists today. This
enables us to re-examine his conclusions in the light of modern knowledge
and to assess the relevance of factors such as meteorology in influencing
death rates in the City of London.

It is a rather interesting exercise to look at the death rate during periods
of exceptionally foggy weather. We can discover the very foggy days by
looking at daily weather diaries, the best of which, surprisingly enough,

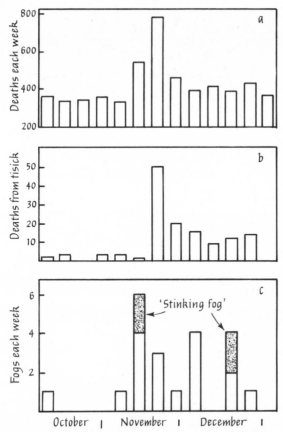

FIGURE 3.12 Death rates each week during severe fogs in late seventeenth-century London: (a) total deaths; (b) deaths from tisick; (c) number of fogs and 'Great Stinking Fogs' (shaded)

does not come from a scientist. The astrologer John Gadbury[37] kept a very complete diary of London weather between the years 1668 and 1689. Just occasionally Gadbury records 'Great Stinking Fogs' in his weather diary. Two of these very severe fogs were experienced in the middle of November 1679 which also happens to have been a week with a very high death rate. The weekly death rate is shown in the upper panel of Fig. 3.12. The actual cause of death can be established by reference to The Bills of Mortality so in the panel below we can show deaths from 'tisick', a lung disease that the bills frequently class with asthma. The bottom panel shows the number of fogs Gadbury observed each week. 'Great stinking fogs' are marked by shaded areas on each of the bars of the histogram. We can see that the total number of deaths increased dramatically after the mid-November week of bad fogs. In addition to this there was a sharp rise in tisick. It is also known from the bills that deaths among old people peaked in the same week. The other notable feature in the figure is the fact that there was a second week of bad fog shortly afterwards. In this week the change in death rate was in no way so marked. It is possible that the most susceptible part of the population was killed off in the first episode and the second found a relatively healthier group of Londoners, so that the death rate did not increase so much.

In essence we find that London in the seventeenth century had begun to suffer from many of the air pollution problems we recognize today. The century also saw the growth of scientific knowledge and London was fortunate in possessing a group of men who showed a particularly clear grasp of the risks air pollution posed for the well-being of the city. Ironically, no use was made of their knowledge.

Notes

1. Dugdale, Sir W. (1658) *History of St. Pauls in London from its Foundation*, Tho. Warren, London.

2. King James I, *A Counterblast to Tobacco*, reprinted by Rodale Press, Emmaus, Penn. (1954). Also Sylvester, J. (1616) *Tobacco Battered and Pipes Shattered*; Davies, Sir J. (1590) *Epigrams and Elegies* and Boas, F. S. (ed.) (1935) *The Diary of Thomas Crosfield*, entry for 13 Aug. 1627, Oxford University Press.

3. Evelyn, J. (1661) *Fumifugium, or The Inconveniencie of the Aer and Smoak of London Dissipated* . . ., printed by W. Godbid for Gabriel Bedel and Thomas Collins, London. The great classic was probably first printed in September (see Evelyn's *Diary* (note 12, below) for September 1661). A second issue of 1661 has the words 'Published by His Majesties Command' deleted. B. White was to print an edition in 1772. It reappeared in editions published by the Swan Press in 1929 and by the Ashmolean Museum in 1930. The National Smoke Abatement Society reprinted it in 1932 and celebrated the work's tercentenary with another reprint in 1961. An important American edition, in *The Smoke of London: Two Prophecies*, selected by J. P. Lodge, was published by Maxwell Reprint Company in 1970. A recent English edition is Evelyn, J., *Fumifugium*, The Rota, University of Essex, Colchester (1976).

4. Evelyn, J. (1664) *Sylva, or a Discourse of Forest-Trees* . . ., London.

5. Platt, H. (1603) *A New Cheape and Delicate Fire of Cole-Balles* . . ., London.

6. Chambers, L. A. (1968) 'Classification and extent of air pollution problems', in Stern, A. C., *Air Pollution*, Academic Press, New York, vol. I, 1–21.

7. Wood is typically low in sulphur compounds. Still there is much worry about emissions of a range of compounds from residential wood combustion: see Cooper, J. A. and Malek, D. (1982) *Residential Solid Fuels*, Oregon Graduate Centre, Beaverton, OR.

8. Manuscripts of the House of Lords, vol. XI; 382–5 from 'Original papers in parliament att the tryall of the Archbishop of Canterbury'.

9. ibid.; *Historical Manuscripts Commission*, vol. IV, 54, House of Lords Calendar 16 Feb 1640–1; *Cal. State Papers (Dom.)*, Charles I. vol. CCCXI (15 Jan. 1636).

10. *Cal. State Papers (Dom.)*, 1547–80, 612.

11. *Royal Commission on Historical Manuscripts*. Third Report, 1872, 28.
12. de Beer, E. S. (ed.) (1956) *The Diary of John Evelyn*, Clarendon Press, Oxford, entry for 7 Nov. 1651. All references to the diary henceforth are from this edition. The dates of the entries are given.
13. Digby, Sir K. (1658) *A Late Discourse . . .*; said to be translated from a French edition by R. White, but the French edition is not known.
14. Maddison, R. E. W. (1969) *The Life of the Honourable Robert Boyle FRS*, Taylor & Francis, London.
15. Grant, D. (1957) *Margaret the First*, Hart-Davis, London.
16. Cavendish, M. (1653) *Poems and Fancies*, J. Martin & J. Allestrye, London.
17. *Letters and Poems in Honour of the Incomparable Princess, Margaret Duchess of Newcastle*, J. Martin & J. Allestrye, London (1676).
18. Petty, Sir W. (1690) *Several Essays in Political Arithmetick*, R. Clavel & H. Mortlock, London. Petty certainly expended considerable effort in gathering coal consumption figures. He collected those for Dublin at the request of Sir John Lowther (representative for the Shire of Westmorland) in 1677; see Landsdowne, Marquis of (ed.) (1928) *The Petty Southwell Correspondence 1676–1687*, Constable, London.
19. Evelyn, J. (1659) *A Character of England* – published a year after Digby's *Discourse on Sympathetic Powder*.
20. See Parry, G. (1976) notes to the preface of the Rota Edition of *Fumifugium*, University of Essex, Colchester.
21. de Beer, E. S. (ed.) (1938) *London Revived*, Clarendon Press, Oxford.
22. Evelyn, J., *Diary*, 4 September 1666.
23. *Cal. State Papers (Dom.)* Charles II, vol. CLXXV, 13 September 1666.
24. Graunt, J. (1662) *Natural and Political Observations . . . Made upon the Bills of Mortality*, London. The edition used here was the reprint in Graunt, J. and King, G. (1973) *The Earliest Classics*, Gregg Int. Pub. It is curious to note that Campbell (in Graunt's biography in *The Dictionary of National Biography*, Oxford University Press (1917 . . .)) states that the *Observations* appeared in 1661, which would make them contemporary with *Fumifugium*. Evelyn had certainly seen Graunt's work by 1666 because it is referred to in his *Londinium Redivivum* (de Beer, E. S. (ed.) (1938) *London Revived*, Clarendon Press, Oxford). De Beer notes that fifty copies of Graunt's book had been distributed to Fellows of the Royal Society, but it seems hard to believe that Evelyn could have seen a copy before *Fumifugium* was published.
25. Spratt, T. (1667) *The History of the Royal Society*, J.R. for J. Martyn, London.
26. Biography of Glisson in *The Dictionary of National Biography*, Oxford University Press (1917 . . .).
27. Burland, C. (1918) 'A historical case of rickets: being an account of the medical examination of the remains of Princess Elizabeth, daughter of King Charles I, who died at Carisbrook Castle, September 8, 1650', *Practitioner*, 100, 391–5.
28. Bentham, H. R., in Robson-Scott, W. D. (1952) *German Travellers in England*, Basil Blackwell, Oxford.
29. Howe, G. M. (1972) *Man, Environment and Disease in Britain*, David & Charles, Newton Abbot.
30. Eddy, T. P. (1974) 'Coal smoke and mortality of the elderly', *Nature*, 251, 136–8.

31. A report with references on rickets in Britain today is to be found in *New Scientist*, 72 (1976), 654.

32. Manley, G. (1952) 'Weather and disease: some eighteenth-century contributions to observational meteorology', *Notes and Records of the Royal Society*, 9, 300.

33. Henshaw, N. (1664) *Aero-chalinos*, Dublin.

34. Boyle, R. (1692) *A General History of the Air* . . ., A. & J. Churchill, London.

35. Hooke, R. (1726) *Philosophical Experiments and Observations*, W. Derham, London; and also Hall, J. (1750) 'On noxious and salutiferous fumes', *Gentleman's Magazine* 20, 454.

36. Justel, H. (1686–7) 'An account of an engine . . .', *Phil. Trans.*, 16, 78.

37. Gadbury, J. (1691) *Nauticum Astrologicum*, London. Gadbury was embroiled in Papist plots and not always entirely appreciated, as the publication *The Character of the Quack-Astrologer, John Gadbury* (1673) suggests.

4

The effect of air pollution in coal-burning London

The changes of the seventeenth century, when coal replaced wood as the major domestic fuel, were more startling than those which took place in the century that followed. Of course the quantity of coal used grew with the population of the city; in fact it grew slightly faster than the population and towards the end of the eighteenth century coal imports exceeded the level of a million tons per annum. The increases in per capita consumption of the fuel reflected increasing wealth and technology. Parallel with the steady increase in coal usage was the adoption of the fuel in nearly all industrial processes. It is easy to believe that along with this increase in fuel consumption came an increase in air pollution, although the rate of increase in the pollution problem was probably enhanced by the fact that London did not expand proportionally to accommodate its increasing population. Thus the fuel was being burnt within the city at an increasing density.

The high levels of air pollutants in the atmosphere of London began to alter the lives of its inhabitants in quite subtle ways. The satires of the poets, Swift and Gay, suggest that walking in the streets of London in the early eighteenth century was not a pleasurable experience.[1] One stood a good chance of being doused in a soot-laden shower of rain or engulfed by an obnoxious mist. In addition to this inconvenient aspect of the urban atmosphere, the streets were dirty with the sootfall. The problem of keeping clothes clean in a smoky city had been noted by Digby[2] in the 1650s, but it had become so serious by the eighteenth century that ladies began to wear cast-iron pattens on their shoes in order to prevent the hems of their dresses becoming coated with the dirt and soot that lay upon the ground, and clothes became so badly 'smoked' that thriving businesses were running in eighteenth-century London to refurbish them.[3]

Preventing the emission of smoke would seem to us to be the best response to the problems, but the cures applied in the eighteenth century were rather different. Jonah Hanway, the famous philanthropist, who is remembered for his concern for the health of the chimney-sweeps, became an object of ridicule because he carried an umbrella about the streets of London. The habit was considered quite un-English at the time;[4] the proper thing to do was to take a sedan chair rather than to ward off the inky rain with such a contrivance. The use of the umbrella in this way might also explain why it is traditionally coloured black. The influence of pollution on fashion was also evident in the nineteenth century: when the American lecturer Ralph Waldo Emerson (1803–82) toured smoke-begrimed Britain, he was told that the hopelessness of keeping clothes white lead to a rather dowdy style of dress among the ladies.[5] Continental visitors noted that the English favoured cream rather than white and certainly there are numerous examples of off-white clothing in Edwardian times. It is tempting to see evidence for this lasting into the present century in the drab clothing that is still a feature of our image of the industrial Midlands. Such a vision is to be seen, for example, in paintings by artists such as Lowry.

Fashion in furnishings was also affected. As far back as the beginning of the 1500s there was occasional note of damage to interiors by coal smoke. In 1510 the Earl of Northumberland ordered a special load of wood at Christmas-time, because coal could not be burnt then for fear that it would damage the arras that was to be hung for the festivities.[6] From descriptions in the seventeenth century we know that hangings and paintings suffered no less in John Evelyn's time. The damage smoke caused to domestic interiors seems to have had an important effect on the style of interior decoration by the eighteenth century.[7] In 1725 the French Ambassador reported that hangings were quite uncommon in London, because they would be rapidly ruined by the coal smoke. In addition to hangings, paper and leather were attacked; books in particular suffered. At the beginning of the nineteenth century we hear of a most notable incident when the problem proved so serious that the famous scientist Michael Faraday wrote a pamphlet about the state of the leather chairs at his London club.[8] The dark-coloured wallpapers found in Victorian homes may also attest to the difficulty of keeping interiors clean in smoky atmospheres.

The damage to the exterior of buildings that had attracted James I's attention in earlier centuries continued unabated. Old St Paul's had burnt down during the Great Fire of London and the new building was still seen as one of the principal victims of the corroding effect of coal smoke. Ironically the rebuilding was partly financed by a tax levied on the coal imported into London. There is no evidence that this was an environmental move, as it sometimes is in the present century, to oblige the coal users to pay for a kind of licence to pollute. Londoners and visitors to the city during the early eighteenth century described the cathedral's sorry state, even before its completion. Timothy Nourse in *An Essay of the Fuel of London* tells us that the stones of many elegant buildings were eaten away, peeled and flayed to the very bone.[9] In addition to St Paul's he lists as badly corroded: St Peter's in Westminster, the buildings in the Strand such as

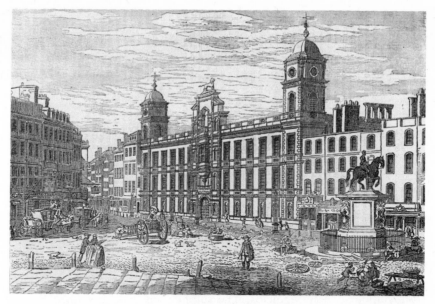

FIGURE 4.1 Northumberland House, one of the buildings that Timothy Nourse mentioned as being badly corroded by coal smoke at the end of the seventeenth century

Sommerset House, the Savoy, the New Exchange, Northumberland House and the ancient buildings of Whitehall.

More modest dwellings also suffered. Some eighteenth-century leases to the houses in fashionable districts in London contained clauses to ensure that their exteriors were repainted triennially, in an attempt to combat smoke disfigurement.[10] The influence of coal smoke persisted until recent times, for early in the present century city houses were painted in dull colours to avoid showing the dirt. The only bright patches in the urban environment were the rapidly replaced advertising posters, and even these faded rapidly because of the pollutant gases in the air.[11]

The corroding action of pollutants

There appears no doubt that the pollutants in London's atmosphere had a detrimental effect on the very materials from which the city was built. The corrosion of building materials can be understood by examining the chemistry of the processes involved. In this section the mechanism of attack on building materials will be considered, with the benefit of 200 years' hindsight. The reader not so interested in such detail could easily skip to the next section.

TABLE 4.1 Sulphur content of coals

Origin	Type	% Sulphur
Notts. slack	bituminous	0.45
Lancashire	bituminous	1.38
Yorkshire	bituminous	1.20
Durham	bituminous	1.00
Scotland	anthracite	0.10
S. Wales	anthracite	about 0.7
E. Indian	semi-bituminous	2.50

Source: Bone, W.A. (1918) *Coal and its Scientific Uses*, Longman, Green, London.

The early members of the Royal Society who wrote on pollution considered the corrosive nature of the coal smoke to arise from some kind of volatile acid. Evelyn thought it to be a sulphurous compound. These suggestions were certainly along the right lines, because we know today that much of the damage in the atmospheres of coal-burning cities arises from sulphur-containing acids. The sulphur which is present in coal as compounds such as iron pyrites (FeS_2) and organic sulphides is oxidized to the volatile sulphur dioxide (SO_2) during combustion. This is emitted as a gas which is invisible but will accompany the visible smoke at least for a short distance. In high concentrations it has a sharp sulphurous odour and can be tasted in the air; hence the complaints of Queen Elizabeth I about the annoying taste of sulphur in the air of Westminster.[12] The sulphur comprises only a very small percentage of the total composition of coals, the softer bituminous coals generally containing the largest concentrations, as can be seen from Table 4.1. Note the low sulphur content of Scottish coal. This no doubt contributed to its early acceptance as a fuel in the north.

Interestingly, when considering damage to materials in the environment, sulphur dioxide is probably not the main culprit. Most of the damage occurs through the effects of a secondary pollutant, sulphuric acid. This acid is produced through the oxidation of the sulphur dioxide. In urban environments this oxidation can take place in polluted fog droplets or when the gas is absorbed onto the surface of materials. The process occurs most rapidly under conditions of high humidity. The comparative ease with which sulphuric acid is formed in a coal-burning city means that ultimately it is sulphuric acid and not the primary pollutant sulphur dioxide which is responsible for corroding the urban fabric.

The attack on the softer limestones is particularly evident and was probably the subject of one of the earliest scientific investigations of air pollution damage. This study was undertaken in the last century by Dr Augustus Voelcker, a chemist retained by the Royal Agricultural Society of England. The results of his work on the corrosion of building stones were communicated to the Society of Arts in 1863.[13] His report described the corrosion of a church built of limestone (calcium carbonate). Although the stone was of the highest purity, year by year a thick dark-coloured crust built up on the stone. The crust was highly soluble and proved to be more than 50 per cent calcium sulphate. This showed that the polluted conditions had converted the calcium carbonate into calcium sulphate.

At first it may seem surprising that this process causes so much damage. After all it merely converts limestone (calcium carbonate) into gypsum (calcium sulphate), another mineral quite familiar in buildings and better known as plaster of Paris. However, two properties of gypsum encourage rapid deterioration of stonework. The molecular volume of gypsum is considerably greater than that of the limestone. This means that the calcium carbonate expands as it is converted to calcium sulphate so the pollution-induced transformation of limestone into gypsum is accompanied by severe mechanical stresses. These serve to disrupt the limestone. The second important property of gypsum is that it is much more soluble in water than calcium carbonate. This means that once the encrustation forms it is quite rapidly weathered by rainwater.

Modern studies of very old building stones from the colleges at Eton have shown that the pollutant sulphur has penetrated about 5 mm into the stone. In newer stones of similar composition this has taken place to a much smaller extent, but then they have been exposed for much shorter periods. If one considers that rate of attack then the modern stones in urban environments appear to be corroded at much higher rates than the older stones in cleaner environments.[14]

Metals, too, are attacked by pollutants derived from coal-burning. John Evelyn suggested in the seventeenth century that the rate of corrosion of iron was at least a hundred times slower in the country than in London.[15] Modern studies continue to show that the rate of corrosion of metals is greater in urban environments than in their rural counterparts. As in the weathering of building stone, the corrosion of metals is enhanced by the sulphuric acid that is produced by oxidation of the pollutant sulphur dioxide. Glass may also be attacked by polluted air. Modern glass is a remarkably durable material, but medieval glass which contained a rather large amount of calcium was more susceptible to attack by both water and sulphur dioxide.[16]

As with outdoor materials, the damage to materials indoors results largely from the acidic nature of sulphur dioxide and its oxidation products. The damage to paper has been extensively studied in recent times because of the large amount of damage that has occurred to books and documents whose safe storage is of great importance. The attack occurs with greatest severity around the edges and is characterized by a yellowing and loss of mechanical strength in the paper. It was once thought that the rate of absorption of sulphur dioxide was increased with humidity, so a very dry atmosphere would favour preservation of paper. However, the rate of uptake of sulphur dioxide by paper does not seem to vary much with humidity, and the loss of water by the paper at low humidities may actually accelerate the damage.

Wallpaper is also exposed to damage by atmospheric pollution, although there are few wallpapers that anybody could tolerate long enough for sulphur dioxide damage to be a problem! Increased damage to wallpaper may occur at points of contact, where sweat deposits encourage enhanced deposition of sulphur dioxide. However, in general the discoloration due to the deposition of suspended soot onto walls has probably been more objectionable than the loss of mechanical strength and more gradual deterioration brought about by sulphur dioxide.

City gardens

The damaging effect of air pollutants on plant life was recognized in medieval times. As we have seen, the civic legislators of the fifteenth century took steps to prevent damage to the fruit trees near Beverley in Yorkshire which suffered from air pollution generated in brick-making operations.[17] By Jacobean times, damage to plants by air pollution was evidently well accepted. Platt suggested that the coal-balls he formulated would give off smoke that was less damaging to the gardens of the noblemen of London than the coal ordinarily used.[18] John Evelyn, whose horticultural interests have already been mentioned, made several comments on the problem. We learn that the anemone, which grew in the London of his youth, had quite disappeared by the time he was writing *Fumifugium*. It is hardly a plant that we would even consider common in London today, although there are reports of it occurring on Hampstead Heath in recent times. During coal shortages Evelyn noted that the gardens and orchards of central London bore plentiful quantities of fruit. The Restoration brought an end to these fuel shortages and by the end of the seventeenth century the atmosphere of London was so bad that many types of plants became difficult to grow in the city.

At the beginning of the eighteenth century Thomas Fairchild wrote *The City Gardener*, a book which comes at the head of a long catalogue of works on city gardening that continue unabated to the present day. These books attempt to advise the city dweller on the best methods to use in urban areas. It seems that Fairchild was persuaded to write the book 'so that everybody in London or other cities where coal was burnt might delight themselves in the pleasures of gardening'. As with many books which give advice about urban cultivation, it is mostly concerned with the types of plants that are hardy enough to survive in the city air and soil. A noticeable characteristic of this book, and of those that have followed it, is a lack of anything but a passive response to the problems of air pollution.[19]

Later in the eighteenth century the situation must have become even worse than in the days of which Fairchild wrote. According to Thomas Gray, the author of *Elegy Written in a Country Churchyard*, the very lime trees which Fairchild thought could grow in London's sour atmosphere were losing their leaves because of the soot.[20] B. White, the editor of the only edition of *Fumifugium* to be published between Evelyn's time and our own, was amused at John Evelyn's comments on the extent of London's vegetation in Restoration times. White concluded that it had become much more limited with the passage of a hundred years.

When discussing damage to buildings we saw that it was not the sulphur dioxide arising directly from burning coal that caused most of the damage. Much of the corrosion was caused by its oxidation product, sulphuric acid. Of course the acid can damage plants when in moderate concentrations and leach nutrient elements from plant leaves and soil. But in general plants and most English soils are relatively resistant to low concentrations of acid in rainfall. It does seem that vegetation is injured by gaseous sulphur dioxide directly. At very low sulphur dioxide concentrations the plant leaf may be able to respond to the presence of the absorbed gas by converting it to sulphate which is an important plant nutrient. However at higher concentrations the leaf can no longer cope with the amount of pollutant, the cells of the mesophyll collapse and the leaf takes on a dull green water-soaked look. On drying, these damaged areas appear whitish. Sulphur dioxide is usually accompanied by smoke in European cities, and this also contributes to plant damage by covering the leaf, thus lowering the transmission of light through to the chloroplasts. Smoke can also block the stomata which allow the transfer of gases in and out of the leaf.[21]

Meteorology and the dispersal transport of air pollutants

Cities change the composition of the air within them, but if there is enough pollution its effects will be found in the country. We have already talked about the fact that medieval and renaissance industry tended to be located in the forests. There is little comment on air pollution there, although just occasionally sensitive accounts of the environmental changes wrought by forest-based industries can be found.

GLYN CYNON WOOD

Aberdare, Llanwynno through,
all Merthyr to Llanfabon;
there was never a more disastrous thing
then the cutting of Glyn Cynon.

They cut down many a palour pure
where youth and manhood meet;
in those days of the regular star
Glyn Cynon's woods were sweet.

. . . Many a birch-tree green of cloak
(I'd like to choke the Saxon!)
is now a flaming heap of fire
where iron workers blacken.

For cutting the branch and bearing away
the wild birds' habitation
may misfortune quickly reach
Rowenna's, treacherous children!

Rather should the English be
strung up beneath the seas,
keeping painful house in hell
than felling Cynon's trees.[22]

(anon.)

This is a wonderful piece of early environmental poetry. It is interesting to see how it is tinged by the political context of exploitation of natural resources by a foreign power.

The first example of long-range transport of air pollutants from the British Isles comes from moorlands rather than the forests. Through the fifteenth century[23] there were numerous acts passed against 'muirburning'. These acts are Scottish, so it is not surprising to find that an English act concerning moor-burning was first passed under James I. The purpose of these acts is not entirely clear, although they may have arisen from a desire to prevent large and uncontrolled fires. However, John Evelyn claims that the English acts had an element of environmental concern in their formulation, because in Elizabethan times, it was claimed, large-scale burning of moorland ruined the French vine crops in the bud. A southerly drift of pollution would have been more likely in those times than at present because the pollution would have drifted southwards under the pattern of atmospheric circulation that prevailed during the Little Ice Age.[24] These episodes must be some of the earliest trans-national pollution incidents. Today different circulation patterns mean that we might more readily expect to find our pollution ending up in Scandinavia. It is unlikely that the pollution would have been severe enough to cause damage to vines in France, but it is likely that the prevailing winds carried soot that far.[25] Soot may have been observed and it is probable that the smoke could have been smelt, so the French wine growers felt that they had good grounds for their complaint.

In the middle of the seventeenth century we find the beginnings of the modern science of meteorology, so it is not surprising that there is a large amount of quite accurate weather information available for the century that followed. Naturally, the effects of pollution did not escape the attention of the earliest of observers. 'Great Stinking Fog' appears in meteorological registers and observers were warned of the increasing vigilance necessary to distinguish drifting town smoke from fog.[26] The smoke from London must have spread out over vast areas to the lee of the city as an urban plume. This plume would be similar to the smaller plumes that can be seen coming from single-point sources, such as factory chimneys, except that the urban source is spread out over a large area so that the plume is more diffuse than the ones we are more likely to observe, although they are easily seen in satellite pictures of the Earth. Large modern cities may have very extensive urban plumes which can be detected hundreds of kilometres downwind, at which point they may be more than 50 km wide.[27] Scientific studies of urban plumes have begun only quite recently. This seems to have led to a feeling that the urban plumes themselves are relatively recent phenomena. Of course this is not so and urban plumes would have been associated with the earliest of cities, but they were probably too indistinct for the unaided observer to notice.

The earliest record of an urban plume from the city of London is of the plume from the Great Fire in September 1666. The description of this extreme event shows us that urban plumes from the metropolis were unfamiliar to even the most careful seventeenth-century observers under normal conditions.[28] During the fire, smoke caused the sunlight in London to be quite red, and as far away as Oxford the philosopher John Locke was able to note in his diary:

> Dim reddish sunshine. . . . This unusual colour of the Air without cloud made the Sunbeams of a strange red dim light, was very remarkable. We had then heard nothing of the Fire of *London*. But it appeared afterward to be the smoke of London then burning, which being driven this way by an Easterly wind, caused this odd Phenomenon.[29]

It is quite evident from the description that Locke had not noticed the urban plume before, so one must assume that London's urban plume was not normally visible at such distances in the seventeenth century. John Evelyn calculated that the plume from the Great Fire would be nearly 50 miles in length. Sadly he gives no details of the calculation. If he had we might have gained some further insight into his understanding of the meteorological controls on the dispersal of air pollutants.

We know some of Evelyn's thoughts on the role of climate in controlling pollution from *Fumifugium*. It is clear from this book that he appreciated the importance of stable conditions that allowed the accumulation of high levels of atmospheric pollutants. The question of wind direction also receives attention, particularly the southerly winds which drove pollution from the industries near Lambeth across the river to the city of London. In *A Character of England* he notes the predominance of fine houses to the west of London, built there so that the nobles could avoid the stink of the eastern parts of the city. Evelyn himself lived in High Holborn for some time, which suggests that he was not averse to using a favourable situation to escape from the fumes of the city.

In the eighteenth century London's urban plume became more readily observable. The 'dingy, smoky' appearance of the air in dry weather was noted at Selbourne in Hampshire by the famous naturalist Gilbert White.[30] He made regular observations of the phenomenon and agreed with the opinion of local inhabitants as to its origin. The local people said that they could smell the 'London Smoke' and the characteristic smell was enough to prove that the metropolis was indeed its source. White noticed that the smoke was always observed during periods of steady easterly winds and even suggested that one could check whether the city was a source by making observations to the east of London where the smoke should be absent. The easterly winds not only helped carry the London Smoke down into Hampshire, where White made his observations, but the clear stable conditions associated with such an easterly flow would lead to a fairly restricted dispersion of the urban plume.

Relatively fine conditions are required if the urban plume is to be observed at long distances. This would have been known to White because some years earlier another English meteorologist, Huxham, living at Plymouth, had drawn attention to the importance of rain in clearing the atmosphere of suspended material.[31] However, sunny weather associated with the stable easterly conditions may then have had a subtler role to play. Such conditions would have aided the generation of secondary pollutants through photochemical reactions that are now well understood from studies of the chemistry of Los Angeles smog. The ease with which a plume may be observed under a given set of lighting conditions is largely determined by the concentration of suspended particulate matter. It is now known that the gaseous pollutants in the plume may react to produce new particles when exposed to sunlight. The secondary particulate material produced in this way adds to the smoke already present in the plume, to some extent counteracting the effect of dilution with distance.

This will considerably enhance the visible length of the plume. It is likely that such photochemical processes also contributed to early observation of London's urban plume. The importance of photochemistry in the chemistry of plumes, as distinct from purely urban air pollution, has only been realized in the last decade or so. In fact it has become rather important today because it is possible that photochemical pollutants over England are derived from sources in Europe. These were a matter of particular concern in the UK as pollution concentrations were very high during the long dry

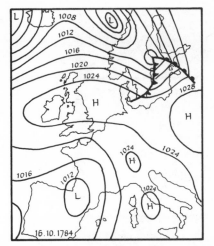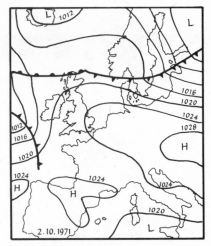

FIGURE 4.2 Weather maps for polluted days in the eighteenth and twentieth centuries

summers of the 1970s. Some scientists fear that the industrial activities of Europe are increasing the concentration of photochemical pollutants over south-east England during stable anticyclonic conditions. Figure 4.2 shows two weather maps for days when there was an easterly drift of pollution over the British Isles. The first was assembled from early meteorological observations by John Kington of the Climatic Research Unit and is for one of the days on which Gilbert White observed London Smoke. It has one striking feature: a broad tongue of high pressure extending back over the British Isles. This feature is also found on the modern map next to it, which shows the situation on a recent day when photochemical pollution was marked in north-western Europe. Blocking situations such as this were rather frequent in the 1780s when White made his observations. Apart from giving the stability and long sunshine hours necessary to generate secondary pollutants, the dry weather and clear skies made conditions for observation of the plume more favourable. Cloudy weather would make observation more difficult and rain would wash the particulates from the air.

Health and trace components of the air

Early meteorological work was largely carried out by physicians whose interest strongly influenced the course of English meteorology. The plagues of the seventeenth century resulted in continuing studies of the miasmatic effect of air on health and led to the production of works such as

FIGURE 4.3 Dr John Arbuthnot, author of a book on the health effects of air pollutants

Henshaw's *Aero-Chalinos*. The matter was re-discussed at regular intervals through the eighteenth century by various medical writers such as Arbuthnot, Mead and Walker.[32] Dr John Arbuthnot (1667–1735) is often remembered as a friend of the poet Alexander Pope and supposed recipient of the well-known *Epistle to Dr. Arbuthnot*. In 1733 Arbuthnot published *An Essay Concerning the Effects of Air on Human Bodies*, which catalogues much of the earlier learning: the corrosive effects of the 'vitriolick salts' on hangings in rooms, the 'unfriendliness' of city air to the lungs due to the 'sulphureous Steams of Fuels', and the high death rate among urban infants.

Richard Mead, Pope's physician, was, like Arbuthnot, treated in a number of epistles (e.g. 'And Books for Mead and Butterflies for Sloane'). With book collecting, writing and scientific work it is surprising that either Mead or Arbuthnot can have ever had much time to spare for their patients. Mead wrote an important book on poisons which included a section on those in the air and contributed notes to a book on ventilating the holds of ships.

A little later Walker wrote *A Philosophical Estimate of the Cause, Effects and*

Cure of Unwholesome Air in Large Cities (1777). He claimed that lungs can actually become so accustomed to the polluted air of the city that they are offended by that of the country. He distinguished, as Sir Kenelm Digby had done, between the emissions of coal burning, which are acid, and the alkalis (e.g. ammonia) emitted by animals or the emissions from putrefying matter. He suggested, from long observations, that the air of London is a little healthier at street level than in the tops of houses. It is interesting to consider that the poorer people may well have occupied these upper levels in the days before lifts.

It is clear that such men directed that asthmatic and consumptive patients be removed to the country so that they might avoid the harmful effects of the city air. The medical poet John Armstrong (1709–79) wrote in the *Art of Preserving Health* (1744) that we should:

> Fly the city, shun its turbid air
> Breathe not its chaos of eternal smoke

This and the note by Swift[33] that doctors in Dublin remove their patients to the suburbs show a continued awareness of the dangers of city air to health, but have little that is fundamentally new to offer. After all, doctors in Imperial Rome had given the same advice to those with weak lungs and money in ancient times. Suggestions for active air pollution control are even further from the minds of these writers. They appear only to consider passive responses.

As chemical knowledge improved in the century that followed the pioneering work of Digby and Evelyn, chemists became keenly interested in the composition of the atmosphere. They continued to be particularly anxious to identify the materials that might be biologically active. The discovery of such materials would lend support to the miasmatic approach to the origins of disease. As we have seen, sulphur was frequently considered to be a harmful element of smoke, even tobacco smoke.[34] Other scientists thought that the contagious components were the trace elements such as lead, antimony and mercury.

There is an interesting connection between these ideas and early studies of carcinogenesis. Evelyn had wondered if it might even be possible for some atmospheric contaminants to pass through the skin. This perceptive thought precedes Percival Pott's work on diseases among London chimney-sweeps by more than a century.[35] Continual contact with organic chemicals in the smoke can induce a variety of diseases. In the mid-eighteenth century John Hall wrote a tract, *Cautions Against the Immoderate use of Snuff* (1761), that represents another very early contribution to the understanding of chemically induced cancers. Significantly, Hall was interested in industrial disease also and wrote about the fumes and diseases around smelters. However, any measurement of the concentration of trace metals or organic carcinogens in air remained quite beyond the limited analytical capability of the time. Even today the minute quantities of these materials in the atmosphere taxes modern analytical techniques; little wonder that it was not possible for the early chemists to characterize the presence of miasmatic compounds in the atmosphere.

Early analyses of the atmosphere

The above is not to say that the efforts of the early chemists were without importance. As has already been mentioned in the previous chapter, Robert Boyle at the end of the seventeenth century tried to devise sensitive techniques for the estimation of various components in air.[36] Several methods for doing these analyses were suggested. He noticed that the discoloration of metals depended to some extent on the trace components in the air. The fact that corrosion on a house in Buckinghamshire was not visibly different in any direction suggested to him that the distribution of the trace components was isotropic with respect to wind direction. As a more specific analytical technique Boyle advised the experimenter to 'hang up clothes or silks died with colours . . .' and to note any fading or changes in colour. Boyle thought this would indicate the presence of particular nitrous or salino-sulphureous spirits.

The technique of using the fading rates of dyes would probably have had a reasonable chance of working in very polluted atmospheres. Very close to pollution sources, damage to dyed cloths was a serious problem. In the 1680s smoke from a glass-house belonging to Henry Tindall was alleged, by people drying cloth near by, to be a potential source of damage to the 'colours of their Cloathes and Stuffes'.[37] In fact as an analytical technique the principle is still in use in some methods today. Less than a hundred years ago, Witz, a Parisian scientist, noticed that the lead-based paints on posters in polluted urban air faded very rapidly. This observation allowed the development of methods for determining sulphur dioxide concentration in the air, using lead salts.

It is surprising how early some trace gases were detected in the air.[38] In 1716 the Italian B. Ramazzini (1633–1714), who is often thought of as the father of industrial medicine, detected nitrogen dioxide in air. The French chemist Rouelle found sulphur dioxide, or at least sulphate, by modifying the method outlined by Boyle. In 1744 he exposed cloths soaked in lye, probably a mixture of potassium and sodium hydroxides, to the air and found sulphate present at the end of the experiment. Scheele, the Swedish chemist best remembered for discovering oxygen, observed encrustations of the ammonium salts on the tops of bottles containing acids. This observation lead to the realization that ammonia was present in the atmosphere. All these early discoveries merely suggested the gases were present in the atmosphere. Often many years had to pass before analytical

FIGURE 4.4 Robert Boyle, the 'skeptical chymist', decided that air was a mixture and not a compound

techniques became sensitive enough to allow chemists to determine the actual concentration of trace gases in the air. Almost invariably the early analyses indicated the gases to be present at a far higher concentration than we know to be correct today. For example, early determinations of ammonia which Scheele showed to be present in the atmosphere were undertaken by H. T. Brown. He estimated that ammonia was present in the air at concentrations of about 6 parts per million, more than a thousand times what we now know the concentration to be.

However, the chemists were able to perform reasonably good quantitative analyses for the major components of the atmosphere, such as oxygen, in the late eighteenth century. Priestley and Scheele independently discovered oxygen and were able to show, for the first time, that air consisted of two major components, dephlogisticated air (i.e. oxygen) and phlogisticated air (i.e. nitrogen). It was readily observed that dephlogiston (i.e. the oxygen) rekindled smouldering tapers and supported life. The phlogiston theory was soon overthrown and more modern ideas of oxygen and nitrogen accepted in its place. However, the importance of the observation that oxygen supported life was not lost at a time when scientists were so keenly interested in the importance of the air to health.

Shortly after the discovery of oxygen and its life-supporting properties, methods were developed for measuring its concentration in air and became very popular.[39] The instruments, known as eudiometers, designed to measure the 'goodness' of the air at various locations, became readily available. Much to the delight of seaside hotel proprietors, the scientist Ingen House proclaimed that the air at the sea coast, with high concentrations of oxygen, was much better than ordinary air. Others tried to establish that the air of the mountains was more salubrious than that in the valleys. All this seemed to be based on the idea that oxygen supported life and therefore it was a good thing; it went without saying that you couldn't have too much of a good thing.

The nitric oxide eudiometer determined the oxygen concentration by measuring the contraction after air was mixed with nitric oxide and the resultant nitrogen dioxide dissolved in water. The instrument, which had been developed by Abbé Fontana, was prone to large experimental errors in the hands of the amateur. This gave rise to numerous heated arguments over the relative merits of airs from different places. Thankfully the period of 'eudiometric tourism' was curtailed when the English physicist Henry Cavendish showed how the eudiometer could work accurately. In 1783 Cavendish used the instrument with almost superhuman care to establish that the oxygen concentration in the air was essentially constant, despite widely varying meteorological conditions, over the sixty days on which he made measurements. His experimental expertise is illustrated by the fact that he obtained an ambient oxygen concentration of 20.83 per cent in his measurements, which differs only slightly from the generally accepted modern value of 20.95 per cent.

FIGURE 4.5 Henry Cavendish, the shyest and richest of English physical scientists, determined the oxygen concentration in the air with great accuracy

Cavendish was rather surprised to find no decrease in the oxygen concentration during calm weather because he considered that the enormous fuel consumption of London should have been responsible for an observable change. In modern measurements it is also true that even in the largest cities the decrease in oxygen concentration brought about by the combustion of fuel is barely detectable. The amount of oxygen present in the air is so enormous that changes in oxygen concentration made by burning fuel are almost insignificant. Some environmental writers of the present day have worried about the fact that we burn so much fossil fuel that we are in danger of using all the oxygen up. This would hardly seem likely as the values for oxygen concentration determined by Cavendish two centuries ago are, as we have seen, very little different from those of today. The best measurements available from the present century suggest an unchanging value of around 20.946 per cent. This agrees with our expectations, because if all the known fossil fuel reserves were burnt in a spectacular bonfire the global oxygen level would but momentarily

FIGURE 4.6 John Jeffries, American scientist, collected air samples above London in 1784

decrease to 20.8 per cent. The same decrease in oxygen partial pressure (which is the effective concentration at the lungs) might be experienced by moving some 60 m higher in the atmosphere. These conclusions do not take into account the fact that oxygen is constantly renewed by photo-synthesis. The concern some people seem to have about our running out of oxygen is so unlikely that the fear of its depletion is sometimes called the *non-problem of oxygen*.[40]

The development of ballooning, a few years after Cavendish's classic experiments to determine the constancy of oxygen in the atmosphere, gave the scientist the chance to extend his measurements to the vertical dimension. An American physician, John Jeffries, planned to make the first mete-orological ascent in England with the eccentric balloonist Pierre Blanchard. Blanchard was well known for ejecting unwanted but fare-paying

FIGURE 4.7 The eudiometer used by Cavendish to determine oxygen concentrations

passengers from his balloon to lighten the load. Thus it is evident that the experiments proposed by Jeffries were not without risk. One of the precautions Jeffries took was to strap himself in with a primitive seat-belt! Records of the food he and Blanchard carried also prove that in-flight catering had very promising beginnings. Cavendish learnt of the plans for a meteorological balloon flight and was able to enlist Jeffries's help in obtaining air samples from various altitudes. The samples were collected during the flight by carrying bottles filled with water. At the appropriate moment the water was emptied out and the bottle sealed up, trapping an air sample.[41] The balloon flight was a great success and the samples returned to Cavendish for analysis. Although the work on these samples was completed a few days after the flight Cavendish never published the results. However, with the aid of Cavendish's laboratory notes and Jeffries' diaries of the balloon flight, it is possible to re-examine the work and see that it showed that the oxygen concentration did not vary with altitude (Fig. 4.8). Cavendish's usual reticence about publishing the results meant that he has not been recognized as the first person to establish that the lower atmosphere has an essentially constant oxygen concentration with height. The honour for this discovery is generally awarded to the Frenchman Gay-Lussac or the Russian Sacharov who made profiles some twenty years later.

Perception of pollution

While the scientists monitored the changes in the atmosphere with instruments of increasing sophistication, the poets and philosophers were

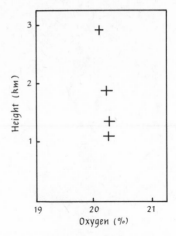

FIGURE 4.8 Profile of oxygen in the atmosphere above London as determined by Cavendish and Jeffries in 1784

conscious that deeper changes were under way. The increasing pressures that man had begun to place on his surroundings were leading to a decline in the quality of the environment. This is expressed most clearly in the poetry of the Romantics, but it is found increasingly in a wide variety of eighteenth-century literary expression. Environmental damage received its most light-hearted treatment at the hands of satirists such as Swift and Gay, who wrote of Londoners surprised by sooty showers, dreary fogs or suffocating mists in their poems *Description of a City Shower* (Swift, 1711) and *Trivia; or Walking the Streets of London* (Gay, 1716). A century later we find Henry Lutrell complaining of the same thing, but at greater length, in his *Advice to Julia* (1820); the environment, it would seem, had not improved.

While the changes in environmental quality that were under way in the eighteenth century were considerable, the changes in the way the environment was perceived may have been even greater. It is the Romantic poets who forced this change upon the human mind and demanded that we look at our surroundings in entirely new ways. This meant that Nature was taken very seriously; poets, such as Blake, were not willing merely to poke fun at the declining condition of their surroundings. They wrote in earnest. Man was being degraded by the condition of his environment. The descriptions of England's dark satanic mills and blackened churches are laden with images of oppression and despair.[42]

The visual artist did not neglect these changes either. Industrial scenes were a popular subject for painters throughout the latter part of the eighteenth century, but a changing mood towards the subject matter is unmistakable. In the earlier paintings, such as those of P. J. de

Loutherbourg (*Coalbrookdale by Night*), the figures within seem to have a noble quality. They toil like heroes amongst the towering chimneys and boilers, which in the half light of furnaces might be mistaken for Grecian columns. By the end of the eighteenth century, the hope had changed to despair: the paintings seem like images of hell and the workmen no longer masters of their own fate. Indeed, the painter John Martin used industrial settings as inspiration for his engravings for *Paradise Lost*.[43]

At the outset many people including artists and philosophers had welcomed the industrial age. The steam engine was hailed as an instrument of social change, which would free men from enormous amounts of unnecessary labour and ultimately eliminate the need for a working class. It hardly seemed possible that an age of despair rather than an age of liberation would follow industrialization, but many reacted strongly to the awful clouds of smoke released by the new engines. The steam-driven pumps that lined the Thames appeared, 'in the rare intervals that they were working, more determined to suffocate the inhabitants of London than to supply them with water'. The pumping engine of the York Water Buildings suffered considerably, both from mechanical inadequacies in its design and from bombings. Sadly the latter were probably not so much by angered environmentalists as by the agents of rival water companies.[44]

A great literary movement does not arise isolated from the developments of its age. The increasing impact of industrial growth and new technology on the environment is an important factor in the development of eighteenth-century thought. Pollution and the human suffering it engendered were forces that influenced the development of the Romantic movement. Two hundred years later, we can now find neo-romantic aspects to the environmentalism of today. At the centre of the Romantic philosophy is a re-emphasis on the importance of nature and the individual, which had been neglected during the rapid technological advance of the early industrial era. Along with this came the desire for simpler, more natural life-styles and the deification of the Earth. These notions have persisted to the present. Today romantic ideals can even be found in some areas of scientific thought. The theme of a Mother Earth has in recent years been discussed in the pages of such reputable scientific journals as *Atmospheric Environment*. The concept has been explored by the scientist James Lovelock under the title of the 'Gaia Hypothesis'.[45]

The change in thought brought about by the re-evaluation of Nature which occurred during the Romantic period can be seen in the evolution of the very word 'pollution'. Once it may have meant 'to wash over' as a river does when it floods its banks, but gradually it assumed a religious connotation concerned more with defilement. Then the poets of the late eighteenth century took the word, laden with all its religious feeling, and applied it to what we now call environmental pollution. One of the earliest examples of this occurs in Anna Seward's poem *Coalbrookdale*:

FIGURE 4.9 The York Water Buildings were situated in a fashionable area of London. The early steam engine provoked much unfavourable comment

> while red the countless fires,
> with number'd flames, bicker on all the hills
> Dark'ning the summer's sun with columns large
> of thick, sulphurous smoke, which spread like palls
> that screen the dead, upon the sylvan robe
> of thy aspiring rocks; *pollute* thy gales
> and stain thy glassy water-sea, in troops
> the dusk artificers, with brazen throats
> swarm on thy cliffs, and clamour in thy glens
> sleepy and wild, ill suited to such guests.[46]

The growing reverence for Nature encouraged by the early Romantics allowed the word 'pollution' more frequently to assume its modern meaning. The words chosen by writers at various times in the past to describe pollution offer clues to how they felt about the human impact on the environment. The Romantic influence led to the wider acceptance of a subtle and almost religious feeling towards the environment. It was evident in writing, where environmental pollution became taboo: an act of impiety. Even in the present century words of spiritual origin are not uncommon in descriptions of environmental damage. D. H. Lawrence, in an account of Tevershall pit bank that can be found in *Lady Chatterley's Lover*,[47] calls coal 'excrement' and the resultant pollution 'black manna from the skies of doom!' Words such as 'incubus', 'ghouls' and 'ghosts' all find their way into writings of the Victorian period about air pollution and seem very different from those such as 'horrid', 'pernicious' or 'corrupt' that would have been chosen by John Evelyn two hundred years earlier.[48]

In the late eighteenth and early nineteenth centuries much of the description of London's air pollution is to be found within the writings of visitors to the city, who appear to be more sensitive to the quality of its air than those who dwell in it. It is not that the Londoner didn't notice it, but as the essayist Charles Lamb said of the city's 'beloved smoke', it was the medium most familiar to him. Dickens was sensitive to the effect that pollution had on human welfare; being a Londoner himself it sometimes seems as if he is more acutely sensitive to it when writing about other cities. In London where he knew that the blackness and gloom seemed to be spread everywhere, he was still able to write quite nostalgically of the pollution at other times, referring to smoke as 'London's ivy'.[49]

Occasionally all the normal responses that we expect to find to air pollution can be completely reversed. In a novel by the Canadian author Sara Jeanette Duncan, written at the end of last century,[50] the smell obscuration and blackening effect of the smoke seem not to provoke disfavour at all:

> There was the smell to begin with . . . always more pronounced in the heart of the City, than in Kensington for instance. It was no special

odour or collection of odours that could be distinguished – it was a rather abstract smell – and yet it gave a kind of solidity and nutriment to the air, and made you feel as if your lungs digested it. There was comfort and support and satisfaction in that smell . . .

I don't know whether you will like our admiring you on account of your griminess, but we do. At home we are so monotonously clean, architecturally, that we can't make any aesthetic pretensions. There is nothing artistic about white brick.

A little of the ambivalence we have seen in writers can be found in the attitude towards pollution among painters. A number of paintings by Vincent Van Gogh illustrate this: in some of his paintings of factories the sky is filled with smoke in a playful way and the paintings are full of bright colours. In others, however, the smoke lends nothing but despair to depressing townscapes.

It would seem that despite all the progress that has been made in the unambiguous identification of toxic materials in the atmosphere through the advances in analytical chemistry, there is more to the pollution of the atmosphere than that. The individuality of the human spirit will never allow anything quite so unambiguous to be said about the perception of pollutants.

Notes

1. Swift, J. (1711) *Miscellanies in Prose and Verse*; Gay, J. (1716) *Trivia, or the Art of Walking the Streets of London*, Bernard Lintott, London. Although these are readily available in modern anthologies, Lutrell, H. (1820) *Advice to Julia*, London, is more difficult to find and to read.
2. Digby, K. (1658) *Discourse on Sympathetic Powder*, London.
3. Kalm, P. (1982) *Kalm's Account of His Visit of England on His Way to America*, transl. J. Lucas, Macmillan Publishers Ltd., London; the refurbishing of clothes is mentioned in Grosley, P. J. (1772) *A Tour of London*, transl. T. Nugent, London, 33 and 72–3.
4. Crawford, T. S. (1970) *A History of the Umbrella*, David & Charles, Newton Abbot.
5 Emerson, R. W. *Journals*, reprinted in Allen, W. (1971) *Transatlantic Crossing*, Heinemann, London; see also Simon, L. (1968) *An American in Regency England*, Robert Maxwell, London.
6. *Duke of Northumberland's Book* (1520).
7. Van Muyden, Madame (ed.) (1902) *A Foreign View of England During the Reigns of George I and II*; a nineteenth-century comment is to be found in Gaskell, E. (1977) *North and South*, Penguin, Harmondsworth, 134.
8. Parker, A. (1955) *The Destructive Effects of Air pollution on Materials*, National Smoke Abatement Society, London.
9. Nourse, T. (1700) *Campania Foelix*, London and also Quarrell, W. H. and Mare, M. (1934) *Travels of Zacharius Conrad von Offenbuch*, Faber & Faber, London.

10. Malcolm, J. P. (1810) *Anecdotes of the Manners and Customs of London During the Eighteenth Century*. It is interesting to consider how polluted the atmosphere might have to be to warrant repainting triennially. On modern acrylic paints a 50 per cent reduction in reflectance would require a three-year exposure at concentrations of about 300 μg m^{-3}; Beloin, N. J. and Maynie. F. H. (1975) 'Soiling of building materials', *J. Air Pollut. Control Assoc.*, 25, 399–403.

11. Marsh, A. (1949) *Smoke*, Faber & Faber, London, 112.

12. *Cal. State Papers (Dom.)*, 1547–80, 612.

13. Voelcker, A. (1864) 'On the injurious effect of smoke on certain building stones and on vegetation', *J. Soc. Arts*, 12 146–51; Geike, A. (1880) 'Rock weathering as illustrated by Edinburgh church yards', *Proc. Royal Soc. Edin.*, 10, 518–32.

14. Brown, R. C. and Wilson, M. J. G. (1970) 'Removal of atmospheric sulphur by building stones', *Atmospheric Environment*, 4, 371–8.

15. Evelyn, J. (1661) *Fumifugium*.

16. Newton, R. (1975) 'The weathering of medieval window glass', *J. Glass Studies*, 17 161–8; and 'Air-pollution, sulphur dioxide and medieval glass', *Corpus Vitreatum Medii Aevi, Newsletter*, 15 (1975) 9–12.

17. Leach, F. (1900) 'Beverley Town documents', *Seldon Society*, 14.

18. Platt, H. (1603) *A New Cheape and Delicate Fire of Cole-Balles . . .*, London.

19. Fairchild, T. (1728) *The City Gardener*, London.

20. Toynbee, P., and Whitby, L. (eds) (1935) *The Correspondence of Thomas Gray*, vol. I, Clarendon Press, Oxford.

21. Unsworth, M. H. and Ormrod, D. P. (1982) *Effects of Gaseous Pollutants on Agriculture and Horticulture*, Butterworths, London is a detailed reference for this matter.

22. Jones, G. (ed.) (1977) *The Oxford Book of Welsh Verse in English*, Oxford University Press.

23. *Ordo-Judiciarie* in *The Acts of the Parliament of Scotland 1124–1707*, I 342/1 (1814): *Scot. Acts. Jas. I*, II 6/1, for the year 1424 (1814); *Exch. Rolls. Scot.* XI. 395 for the year 1499.

24. Evelyn, J., *Fumifugium*, (1661) where the act is reprinted. Reference to the high frequency of easterly winds is to be found in contemporary writings. Verstegan and the astronomer Tycho Brahe are mentioned by Lamb, H. H. (1977) *Climate: Present, Past and Future*, Methuen, London, vol. II, 463.

25. An early report of transport of material as far as Scandinavia can be found in Ibsen's play *Fire* (1865) and in a note by Brøgger, W. C., *Naturen* (1881). The accumulation of soot in moorland sheep, 'moorgrime', is discussed by Gervat, G. P. (1984). *Clouds at Ground Level; Samples from the Southern Pennines*, CEGB TPRD/L/2700/N84. Long-range transport of materials to Scotland in the last century is the subject of Brimblecombe, P., Davies, T. D. and Tranter, M. (1986) 'Nineteenth-century black Scottish showers', *Atmospheric Environment*, 20, 1053–57.

26. Manley, G. (1952) 'Weather and disease: some eighteenth-century contributions to observational meteorology', *Notes and Records of the Royal Society*, 9, 300.

27. White, W. H. *et al.* (1976) 'Formation and transport of secondary air pollutants: ozone and aerosols in the St. Louis urban plume', *Science*, 194, 187–9.

28. Brimblecombe, P. and Wigley, T. M. (1978) 'Early observations of London's urban plume', *Weather*, 33, 215–20.

29. Locke's journal entries may be found in Boyle, R. (1692) *A General History of the Air*, London. Comments on the red sun occur in Vincent, Rev. T. (1667) *God's Terrible Voice in the City*, London.

30. Johnson, W. (ed.) (1970) *Gilbert White's Journals*, David & Charles, Newton Abbot.

31. Huxham, J. (1772) *Observationes de Aëre*, London.

32. Arbuthnot, J. (1733) *An Essay Concerning the Effects of Air on Human Bodies*, London. The line quoted in the text comes from Pope's *Epistle IV*; in *Epistle to Bolinbroke* (in Pope's *Imitations of Horace*) a line runs: 'I'll do what Mead and Cheselden advise.' Mead, R. (1702) *A Mechanical Account of Poisons*, London (1702); Mead, R. and Watson, 'An account of Mr. Sutton's inventions and methods of changing the air in the holds and other parts of ships', in Sutton, S. (1749) *A Historical Account of a New Method for Extracting Foul Air out of Ships*, London; Walker, A. (1777) *A Philosophical Estimate of the Cause, Effects and Cure of Unwholesome Air in Large Cities*, Robson, London.

33. Armstrong, J. (1774) *The Art of Preserving Health*, London; Swift, J. (1729) *Dublin Weekly Journal*. Dublin was not without its early air pollution problems. Two early proclamations are dated 25 June 1634 and 26 July 1665.

34. In Charles Cotton's poem *On Tobacco*, in Beresford, J. (ed.) (1923) *Poems of Charles Cotton*, Richard Cobden-Sanderson, London: 'Sure 'tis the Devil: Oh, I know it that's it, / Fuh! How the sulphur makes me cough and spit!'

35. Evelyn (1661) *Fumifugium*; Pott, P. (1775) *Chirurgical Observations*, London; Hall, J. (1761) *Cautions Against the Immoderate use of Snuff*, London; Hall, J. (1750) 'On noxious and salutiferous fumes', *Gentleman's Magazine*, 20.

36. Boyle, R. (1692) *A General History of the Air*, London. The use of a cloth to absorb atmospheric pollutants for analysis is once more in vogue. The synthetic material 'tak' has proved very useful; e.g. Jones, J., Lewis, G., Orchard, H., Owers, M. and Skelcher, B. (1973) 'The experience of the CEGB in monitoring the environment of its nuclear power stations', *Health Physics*, 24, 619–25.

37. Sewell, J. R. (1977) *The Artillery Ground and Fields in Finsbury*, LTS Publication no. 120, which takes the information from CLRO Misc. MSS. 29.24.

38. Ramazzini, B. (1716) *Opera Omnia* – his works have been published in a modern edition as Wright, W. C. (1940) *Diseases of Workers*, University of Chicago Press; Rouelle, G. F. (1744) *Mem. Acad.* 97; Scheele, C. W. (1780) *Experiments on Air and Fire*. The idea that overestimates were made during early attempts at measuring the concentrations of trace gases in the atmosphere is mentioned in Brimblecombe, P. (1978) 'Historical changes in atmospheric trace components', *Evolution des atmosphères planétaires et climatologie de la terre*, Centre National d'Etudes Spatiales, Toulouse.

39. Brimblecombe, P. (1977) 'The earliest atmospheric profile', *New Scientist*, 67, 364–5.

40. Report of the Study of Critical Environmental Problems, *Man's Impact on the Global Environment*, MIT Press, Cambridge, MA (1970).

41. Jeffries, J. (1786) *A Narrative of Two Aerial Voyages*, London. The analytical data may be found in Thorpe, T. E. (1921) *The Scientific Papers of the Honourable Henry Cavendish*, Cambridge, vol. II.

42. Blake, W. (1789) *Songs of Innocence and of Experience*, London.
43. Feaver, W. (1972) *The Art of John Martin*, Clarendon Press, Oxford.
44. Some of the problems are mentioned in Dickinson, H. W. (1938) *A Short History of the Steam Engine*, Babcock & Wilcox, Cambridge.
45. Articles have appeared both in *New Scientist* (6 Feb. 1975) and *Atmospheric Environment*, 6 (1972), 579, but the subject is more fully treated in Lovelock, J. E. (1979) *Gaia*, Oxford University Press.
46. Scott, W. (1810) *The Poetical Works of Anna Seward*, Edinburgh.
47. Lawrence, D. H. (1928) *Lady Chatterley's Lover*, privately printed, Florence; reprinted Penguin, Harmondsworth (1960).
48 Evelyn, J. (1661) *Fumifugium*, Gabriel Bedel and Thomas Collins, London.
49. Dickens, C. (1852–3) *Bleak House*, Bradbury & Sons, London, (printed in parts).
50. Duncan, S. J. (1891) *An American Girl in London*, Chatto & Windus, London.

5

Smoke abatement

The late seventeenth and early eighteenth centuries show plenty of evidence of interest in the causes and effects of air pollution, but it was hardly an age of environmental activism. With the possible exception of John Evelyn's *Fumifugium*, early works discuss air pollution in a tone of passive acceptance. Anybody who didn't like the state of a city's air was expected to leave or at least move to the suburbs. If your garden was being damaged, you were advised to find hardier varieties to plant. In these early centuries of industrialization, there was no organized movement for smoke abatement. This did not arise until the nineteenth century, when perhaps the confrontation between Industry and Romanticism allowed a new view of man and his surroundings to emerge.

The conditions in late eighteenth-century London were bad, and those of the growing industrial cities of the North an environmental disaster. Sheffield was described as 'black' by Daniel Defoe (1660–1731) in the 1720s in his *Tour of Britain*, and in the same way by William Cobbett (1762–1835) in *Rural Rides* a century later. Even today, this image remains an inherent part of our concept of the industrial Midlands.[1]

As travel became more common, London welcomed large numbers of visitors. They were appalled at the besooted state of the buildings and the lack of clarity in the air. The Americans, more used to conditions found in an untouched continent, were particularly critical.[2] The historian Francis Parkman (1823–93), looking out from St Paul's (claimed to be the dirtiest, gloomiest church yet) declared that all that could be seen were tiled roofs and steeples, half hidden in smoke and mist. And that was in May, a month of clear weather. This was a far cry from the uninterrupted views of unbroken forests in the New World. James Fenimore Cooper (1789–1851), the author of *The Last of the Mohicans*, stood at the same place and decided that the view had little to offer, because, although he liked mists, he had no taste for coal smoke.

The Londoners, though more accustomed to the city, still tried to escape and breathe the fresher air of the country. For the poor, who were confined to the city, the never-ending gloom must have been oppressive, especially in the winter months. The city was tolerated, by both rich and poor,

FIGURE 5.1 Visitors, from the eighteenth century onwards, frequently complained about the smoky London air

because despite all its misery it symbolized the wealth and opportunities of a nation and empire. Byron's comments in *Don Juan* may be particularly indicative of the Londoner's acceptance of the situation, and the feeling that where there was smoke, there was money.

> A mighty mass of brick and smoke, and shipping,
> Dirty and dusky, but as wide as eye
> Could reach, with here and there a sail just skipping
> In sight, then lost amidst the forestry
> Of masts; a wilderness of steeples peeping
> On tiptoe through their sea-coal canopy;
> A huge dun cupola, like a foolscap crown
> On a fool's head – and there is London Town!

> But Juan saw not this: each wreath of smoke
> Appeared to him but as the magic vapour
> Of some alchymic furnace, from whence broke
> The wealth of worlds (a wealth of tax and paper):
> The gloomy clouds, which o'er it [London] as a yoke
> Are bowed, and put the sun out like a taper,
> Were nothing but the natural atmosphere,
> Extremely wholesome, though but rarely clear.[3]

This feeling was not restricted to the smoke billowing from a factory chimney, but was equally applicable to domestic smoke. The Scottish saying 'may your lum [chimney] keep blithley reeking' has remained a popular greeting to the present day.[4] In the South it has been said that the English regarded smoke from their blazing hearths as part of their birthright.[5] This passionate attachment to the hearth as the centre of the home is deeply embedded in folklore. In addition to being a sign of affluence it can also be seen as a symbol of hospitality and warmth.[6] The link between smoke and these attributes may well have been influential in making domestic smoke even more difficult to legislate against than that from industry.

Early solutions

As we have seen, the earliest of medieval legislators were aware of the fact that much of the air pollution problem lay in the fuel. The simple solution was to return to wood as a fuel, and this was a familiar requirement of the earliest regulations. Proclamations insisting that wood be used instead of coal were issued in the thirteenth century to induce the lime burners to return to less objectionable fuels.[7] This approach received its last really serious examination by Timothy Nourse,[8] who 'thought we should attempt to plant forests and properly tend them' so that London could once more be completely supplied with wood. Nothing ever came of these proposals and no doubt it was quite impossible to return to wood even at the beginning of the eighteenth century in London.

Despite this, wood continued to be seen as preferable to coal in the century that followed. Oscar Wilde reported that a Midlands colliery owner once said that 'the one advantage of having coal was that it enabled a gentleman to afford the decency of burning wood on his own hearth'.[9] Even today wood remains a popular domestic fuel. Wood stoves are very fashionable, although their most environmentally conscious advocates have not usually seen them as the complete solution to the fuel crisis.

Dissatisfaction with the safety of nuclear energy, increasing pollution levels and worry about the rise of carbon dioxide in the atmosphere from fossil fuel combustion, have made the utilization of the energy bound up in the biomass seem ever more attractive. In the last few years there has been some interest in energy plantations where it is proposed to grow fuel either directly as 'wood' or as a material that could be readily converted into fuels such as methanol. Furthermore these plantations would also provide attractive recreational areas. Such an energy programme, however, would use vast tracts of land.[10]

So it did not seem, even in the seventeenth century, that it was possible to return to the old fuels. Where reversion to an older fuel was not possible some early smoke abatement proposals suggested a change to new, better and less polluting fuels. Evelyn was keen to see more charring to produce smokeless fuels; either the charring of wood, which he advocated in his book *Sylva*, or the charring of coal which he witnessed at Sir Joseph Winter's house in 1656.[11] Although the charring of wood is a useful technique because it helps to reduce the wood to a manageable form for packing and transport, the attempts at charring coal so common in the seventeenth century never seem to have met with any lasting commercial success. The same may be said of briquette production, which Evelyn would have been acquainted with through Platt's book on coal balls.

One solution which Evelyn seems to have neglected was to use a different kind of coal. This omission is slightly surprising because there is some evidence that Londoners at the time felt that the coal smoke problem had arisen because the better coal was being exported to Europe. While it seems unlikely that good coals were being removed from the domestic market in favour of export, it is true that cleaner coals were available. In the sixteenth century the anthracites of Wales were suggested as a smokeless fuel for London by Owen[12] in his *A History of Pembrokeshire*. Similarly the coals of Scotland, which made much better domestic fuels, would have been a less smoky replacement for the Newcastle coal. The suitability of 'Scots Coal' for the domestic hearth had probably encouraged the use of coal in Scottish homes by the early fourteenth century, long before it was accepted in London. The advantages of Scots Coal were well known in the eighteenth century; a character in one of Sheridan's plays puts it on the fire to impress a guest.[13] Still, it seems that this fuel remained an expensive luxury.

Changing fuels may seem a rather unsophisticated approach to the air pollution problem, but actually it would have worked quite well had it been assiduously applied in eighteenth-century London. We can be fairly sure of this because it is exactly this technique that keeps sulphur dioxide concentrations in London as low as they are today. Presently this is achieved through the use of low-sulphur oils or natural gas. Perhaps this unsophisticated approach suggested by the early proponents of cleaner air is one of the most effective.

Smoke outdoors may have been bad enough, but that at least was a communal problem and everyone suffered, so incentive to overcome the problem remained low. Indoors it was very much the concern of the individual householder. Considerable ingenuity went into the design of chimneys, smokeless grates and stoves, but stoves were always much more popular on the continent than in the British Isles. The French smokeless stove described by Justel in 1686, was described again without modification by the *Gentleman's Magazine* seventy years later.[14] Air pollution technology seems to have been in no particular hurry to advance. There was so little understanding of how to lower smoke emissions during the

FIGURE 5.2 Count Rumford, an American physicist much interested in the design of stoves and fireplaces

combustion of fuels that in the British Isles English engineers had to consult American experts on matters of smoke control.

One of these experts was Count Rumford, or Benjamin Thompson (1753–1814) as he was known in untitled days. Rumford is often remembered for his contributions to the physics of heat and he took considerable interest in the design of stoves and fireplaces. He may also be remembered for the rather adventurous, though not always scrupulously honest, life he led. He married the widow of the French chemist Antoine Lavoisier whom he found such an impossible wife that he thought Lavoisier rather lucky to have escaped to the guillotine. Rumford advocated the use of fire-clay surrounds to retain the heat and required that they be carefully angled to reflect it into the room.[15] The hob was to be abolished and the chimney contracted just above the fireplace. Some of his key writings on the subject of stoves and chimneys appeared in *Essays*

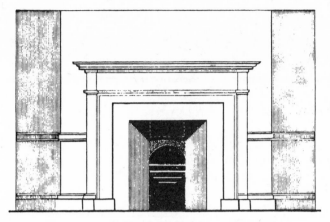

FIGURE 5.3 An older fireplace 'contracted to a Rumford'

Political, Economic and Philosophical in the opening years of the nineteenth century. However, Rumford's stoves and fireplaces had already received some public notice by then. Jane Austen wrote in *Northanger Abbey* that her heroine Catherine Morland, who felt sure she would find gothic horror at Northanger Abbey, discovered that 'The fireplace, where she had expected the ample width and ponderous carving of former times, was contracted to a Rumford, with slabs of plain though handsome marble, and ornaments over it of the prettiest English china'. So for Jane Austen the Rumford stove, it would seem, was a symbol of modernity.

Benjamin Franklin was also very interested in the design of stoves and in fuel economy. In 1766 he advised Mathew Boulton on the matter in relation to the construction of the Boulton & Watt steam engines.[16] Franklin emphasized the need to burn all the smoke and gave two reasons: the first was that the smoke which escapes represents unburnt and therefore wasted fuel, and the second that the smoke is likely to form an insulating crust on the lower surface of the boiler. This crust would be a poor conductor of heat and thus lower the efficiency of the boiler. Franklin's early notions of smoke prevention embody a philosophy that was to remain a central theme of smoke abatement until the present century. Smoke prevention was seen as good practice and sound economy, rather than as a means of preserving the quality of the air. This concept was expressed with the phrase 'burn your own smoke'.

Franklin's smoke abatement technique involved making the smoke from fresh coal pass down through coal that was already thoroughly ignited. This principle was seen, crudely developed, in the stove described by Justel in the seventeenth century, so technically Franklin offered nothing new. However his simple good sense influenced many engineers in the late eighteenth and early nineteenth centuries.

The steam engine

In an odd way, the development of the steam engine had considerable influence on smoke abatement, because it forced engineers to consider the way in which smoke was generated. The concern for the smoke arose because often the public vigorously resisted the new contraptions. As in previous centuries, the charge 'environmental pollution' proved a ready but often ineffective tool to resist changes that were unpopular. The charges were probably accurate and justified when directed against the new steam engines. They were noisy, dirty and dangerous. It is little wonder they were disliked.

While the steam engine allowed protest to focus, it also caused the early environmentalists to take a rather naive approach to atmospheric pollution. For more than a century, their concern continued to be centred on the single-point sources of smoke, notably the steam engines and the large factories. Such an approach was exemplified by the suggestion that there existed a 'Black Catalogue' of the worst polluters.[17]

An interest in large single sources, such as steam engines, drew attention to the problems of industrial chimney heights. This issue had arisen at various times before the nineteenth century. As mentioned in the first chapter, it is possible that there was a minimum height for chimneys in London as early as the fourteenth century. We have no details of these medieval regulations, which did not survive into the industrial period. Smoke from chimneys was regarded as a nuisance and complaints seem to have been dealt with on an individual basis. In 1691 Thomas Legg was able to have a neighbouring baker compelled to construct a chimney 'soe high as to convey the smoake clear of the topps of the houses.'[18] White, in his preface to the 1772 edition of *Fumifugium*, complained of the haphazard design of chimneys and the lack of concern about their heights.

In the early nineteenth century the issue was taken up by Frend[19] in his pamphlet 'Is it possible to free the atmosphere of London from smoke . . .?'. He deplored the lack of any correlation between the emissions from a factory and the height of its chimney. Small factories which had once emitted quite harmless quantities of smoke expanded over the years to become nuisances, because their chimneys remained unchanged. Frend emphasized the need for legislation, but thought that the real solution lay in common sense. This notion has universally characterized almost all the ensuing environmental legislation of the United Kingdom. The nineteenth-century environmental laws arose from liberal thinking which assumed that, given time, people would obviously take the necessary steps that would so clearly lead to better conditions for all. In the case of smoky chimneys, it was thought to be self-evident that there would be great benefits from smoke abatement for both management and public alike.

FIGURE 5.4 A Newcomen steam engine. The smoke from engines such
as these, which consumed vast amounts of coal, particularly offended
Londoners in the late eighteenth century

By the beginning of the nineteenth century there were also signs that the
desire for smoke abatement legislation was not restricted to a few eccentric
enthusiasts. Public administrators began to take an interest in the problem.
Around 1800, the Commissioners of Police in Manchester had appointed a
nuisance committee,[20] which not only recognized the great problem of
smoke from chimneys but had also discovered that there were methods of
solving it. At the very time that Frend's pamphlet was being distributed
among a limited readership in London, the nation's legislators were
discussing the extent to which it would be practical to regulate the use of
steam engines and furnaces so that they would be less prejudicial to the
public health and comfort. The discussion seems very modern when one
considers that they were not only thinking of health but were also

concerned with the less tangible aesthetic notion of 'comfort'. However it is likely that such a vague approach was necessary because there was little hard scientific evidence, at that time, to link air pollution and public health. In a court it would have been difficult to maintain that the smoky atmosphere of London was solely responsible for the health problems experienced by the city. Water-borne disease was probably a far greater, and certainly a far more obvious, hazard to urban health.

The high pressure steam engine designed by James Watt (1736–1819) gradually replaced the more primitive atmospheric engines of Newcomen and Savery. The early Cornish engines took in an enormous charge of coal. It may have been in excess of 30 tons in some of the larger engines but, since they used anthracite, smoke emissions were kept within tolerable bounds, considering also that they were usually sited in fairly remote mining locations. The fact that they were often associated with mines meant that efficient use of coal was of little importance. However, in order to encourage wider use of the newer engines the engineers had to come to grips with the problem of smoke and the fuel losses it implied. They had to prove to their customers that the new engines were an efficient way to generate power. Widespread use also meant that the engine had to be able to use comparatively smoky bituminous coal.

The firing procedure Watt recommended was that the coal should be piled up on the dead-plate; and as there were no fire-doors in the early days there was a small flow of air through the pile of coal that acted as a door. As the coals became hot, the volatiles were drawn into the fire and consumed. Thus the coal on the dead-plate would become partially coked by the time the stoker came to spread it out over the fire and place a fresh load of coal at the entrance to the furance.[21] There is no doubt that this must have been a very good way of reducing the smoke emissions from a steam engine, but required no small degree of skill on the part of the fireman. It was a frequent complaint, in the first half of the nineteenth century, that the men in charge of the furnaces were paid very poorly. This meant that the skilled firemen took other jobs and the management of the furnace fell into the hands of men who were less competent. It should have been obvious to the factory owners that a carefully stoked furnace would lower both fuel consumption and smoke emission, so the skilled stoker would be well worth a high wage.[22]

The original underfired boiler was gradually replaced by one in which the furnace was an internal tube which provided a larger surface area for heating. A greater head of steam could be obtained by spreading the coal throughout the furnace rather than coking it on the dead-plate first. The more rapid combustion process inevitably lead to the production of smoke. This simply added to the problems created by the replacement of the skilled fireman and stokers of the eighteenth century by less skilled and underpaid labourers at the beginning of the nineteenth century. Furthermore the

FIGURE 5.5 Brewing was a smelly, smoky industry from the earliest times

design and construction of furnaces, that had once taxed the skills of the best engineers of an age, was left almost totally in the hands of the brickmaker. It is little wonder that furnaces and boilers became hideously smoky.

In the 1840s, with the art of boiler-making in this gloomy state, Charles Wye Williams started to press for better engines that were smoke-free and more economical.[23] Williams had established a shipping company to run the winter passage across the Irish Sea and was concerned at the unscientific state of boiler design. He didn't believe in Franklin's philosophy of 'burning your own smoke'. Instead, Williams insisted that steam engines should run without producing any smoke. While the conditions required to achieve this were fairly obvious, they were not necessarily easy to maintain. The Royal Navy, whose steamships could not afford to mark their passage with vast clouds of black smoke, had gone to the trouble of making a thorough investigation of coals and had come down in favour of anthracite as a smokeless fuel. In London it was evident that the legislators had also noticed this, as there were cases where individual firms were obliged to use anthracitic coals to lower their

FIGURE 5.6 Local residents forced the brewers Meux and Company Ltd
in Tottenham Court Road, London, to switch to smokeless fuels in the
nineteenth century

emissions of black smoke that had caused a nuisance. For instance, Meux's
Brewery in Tottenham Court Road was forced to use the smokeless fuel
through the pressure of local residents, tired of the damage smoke was
causing to their domestic furnishings. Despite the fact that concern about
smoke abatement was expressed by some administrators at the start of the
nineteenth century, half the century passed with no workable laws reaching
the statute book.

    One of the difficulties may have been that there were few easy solutions
to the pollution problem posed by both the domestic and industrial smoke
in the nineteenth century. The changes of fuel and changes in the location
of industries advocated in earlier times would have been difficult to apply.
The steam engine meant that power could now be produced at the point
where it was required. The waterwheel had offered the possibility of
considerable power in earlier times, but only if you were willing to use the
power near the river. The steam engine allowed power to be generated
within a factory. This brought further industries from the country into the
city. The fuel shortages in Tudor London, discussed in Chapter 2, also had
the effect of inducing some industries to move from the country into the
city, thus increasing air pollution. In a similar way the industrialization of
the city that followed the development of the steam engine also gave rise to
urban air pollution problems. Locating the steam-powered factories in the
cities gave the manufacturers access to an enormous labour market. The

steam engine could be adapted to run on other fuels, but nothing else was available in cities. Relocating industry in the country would have been opposed by the manufacturers.

Nineteenth century legislation[24]

The earliest moves for modern smoke abatement legislation came from M. A. Taylor, the MP for Durham in the 1820s. He found the smoke in Whitehall just as bad as it had been when John Evelyn had complained about the matter a century and a half earlier. However, the steam engine had made town life so undesirable that the rich were vacating the town as rapidly as the poor were moving in, to take advantages of the new opportunities for employment. Taylor had discovered that a Warwickshire furnace owner, Josiah Parkes, had modified a furnace built to the specifications of James Watt. Parkes introduced an auxiliary supply of air and was able to boast that after an hour of running his furnace consumed all its own smoke. A number of Parliamentarians inspected furnaces to which the smoke control devices had been fitted. In the main the results were impressive, although it was clear that low smoke emissions required considerable care in stoking the furnaces. A bill that required furnaces of steam engines to consume their own smoke became law in the early 1820s. It was so weak, however, that it probably had little effect on the air pollution in London.

Significant progress in smoke abatement legislation owes much to W. A. Mackinnon, a Scot who kept up an eight-year campaign to introduce laws to abate smoke. It began in 1843 when Mackinnon chaired a committee whose responsibility it was to inquire into the *Means and Expediency of preventing the Nuisance of Smoke arising from Fires or Furnaces*. The committee met sixteen times and gathered evidence from a number of important scientists, including Michael Faraday, and engineers and manufacturers. The Mackinnon Report failed to persuade the government to take action, so Mackinnon presented a Bill of his own to prohibit the nuisance of smoke from furnaces of 'manufactories'. It applied only to the emissions from furnaces that heated steam boilers, but the bill still ran into trouble through hostile speeches and the introduction of weakening amendments. Finally it was postponed, but even so progress had been made. The public began to understand that smoke was no longer a necessary evil, and the owners of furnaces began to take an interest in smoke abatement. In some ways it is surprising that the owners of furnaces had not acted earlier in attempting to lower their emissions, as they had so much to gain – the smoke escaping from their chimneys represented lost fuel.

The second bill on smoke abatement that Mackinnon brought before Parliament was defeated in 1845, but again not without some victories. Public interest grew further, and *The Times* of 24 April carried an outspoken leader on the subject. Public concern prompted the production of further reports, two by Mackinnon and the De la Beche–Playfair Report. This latter report was written by Sir Henry Thomas De la Beche, director of the Geological Survey, and Sir Lyon Playfair, chemist to the Geological Survey. The report concerned itself with two questions: (i) was the anti-smoke legislation that had been introduced by some towns as part of their Improvement Acts effective, and (ii) was the exemption from smoke controls being requested by many industries, who claimed it was impossible to curb their emissions, justifiable on a technological basis? The report, published in 1846, showed that the local anti-smoke clauses had proved quite unworkable; convictions were near impossible to obtain and fines trivial.

Despite the problems found with existing local smoke abatement legislation, Mackinnon brought a third bill before Parliament in 1846. It was withdrawn, but Mackinnon was under increasing pressure from an aroused public to proceed with his campaign. In 1846 a Public Health Bill incorporated a clause concerning the prevention of smoke. Weak as the clause in this bill was, it did represent a step in the right direction. A similar bill was introduced into the House of Lords, and the City of London promoted a Sanitary Improvement Bill which contained an anti-smoke clause. Once again the industrial interests were able to oppose the planned legislation. They had the clause concerning smoke abatement removed from the Public Health Bill and in the end the other two bills containing smoke abatement measures were also dropped. A fifth bill was introduced to the House in 1849, failing only at the committee stage to become law, and a sixth bill introduced in 1850 failed to get into law under pressure from a well-organized industrial lobby. Thus the seemingly futile struggle by Mackinnon came to an end. It was not, however, without importance because it laid the foundations for the changes that were to take place in the decade that followed, when recovery from a depressed economy and freedom from cholera epidemics allowed the passage of significant reforms.

The next champion of the movement for the abatement of smoke was the Medical Officer of Health for the City of London, John Simon. In his annual report to the Commissioners of Sewers in 1850 he made a strong plea for the abatement of London's smoke, which he maintained was not only unhealthy and wasteful, but also quite preventable. A smoke clause was written into the City of London Sewers Bill which received Royal assent in July of 1851. Unlike the smoke clauses found in previous town Improvement Acts, the London legislation proved satisfactory and in the first year over 115 notices were served on offenders. Simon hoped that clauses that proved successful for the City of London could eventually be widened to cover the suburbs too.

At the end of 1852 Henry John Temple, third Viscount of Palmerston, was appointed to the Home Office. In July of the following year he had a smoke abatement bill prepared and read to the House of Commons. It was considerably modified in its passage, but the fact that it reached the statute book by 20 August was as much a reflection of the effect of Mackinnon's work in the 1840s as it was of Palmerston's energy. One important amendment was the introduction of the phrase 'best practical means' to the wording of the law. This concept has now become embedded in British environmental legislation. Like 'common sense' and 'good practice' these terms have remained part of the flexibility of a system which has been notoriously reluctant to fix quantitative limits to the permissible levels of pollution. The idea was simply that the best practical means available to prevent smoke should be applied.

Palmerston not only pushed the Smoke Nuisance Abatement (Metropolis) Act through the House in 1853, but also followed its enforcement very closely. At first there were few prosecutions under the Act. Then the Home Secretary let it be known that he would not tolerate laxity in the administration of the new laws and a large number of prosecutions followed. It is to be presumed that this had some localized effect on the quality of London's air, but the changes were probably more imagined than real. The earlier legislators realized the importance of monitoring the effects of legislation on air quality as this need was implied in the questions put to De la Beche and Playfair. However at that time the administrators had felt that it was only necessary to monitor the workings of the law. It was but a small step then to require the changes in atmospheric composition to be monitored too. There was no way of establishing the state of the city's air without an air pollution monitoring network, but this had to wait until the First World War to operate on a regular basis.

Public interest

Throughout Mackinnon's long struggles there was a growing public awareness of the air pollution problem. It remained an awareness, however, and did not appear as a strong desire to become involved in the campaign for cleaner air. Interestingly, air pollution did not spring to most people's minds even when they were discussing subjects closely related to smoke and chimneys. Of more than a hundred letters written to a London Iron founder, John Cutler, describing the operation of a smokeless grate he manufactured, only a handful raised the issue of urban air pollution.[25]

Popular journals carried a trickle of articles on smoke prevention throughout the 1850s. *Chambers Journal*[26] was particularly active, leaving its readers in no doubt about the very positive benefits that would accrue from smoke abatement. These were not mere hollow words, either; the publishers had made their printers install a smoke prevention device and

FIGURE 5.7 Domestic chimneys, although not the subject of nineteenth-century legislation, contributed substantially to the pollution of London's air

were careful to point out the great savings that followed from burning their own smoke. Using *Juke's Patent Smoke-consuming Apparatus*, they were able to reduce their annual consumption of coal from about 284 to 264 tons per year, in spite of an increase in output from the presses. In terms of the number of sheets of paper Chambers printed they reckoned on a reduction of more than 20 per cent in coal consumption. This experience seems to have been typical.

The public and the press certainly viewed the industrial lobbies as the villains: not only had they failed to adopt sensible economic practices in their factories but they had fought to prevent the passage of legislation. *Chambers Journal* noted that the iron-smelters and distillers wanted to be exempted from the Smoke Prevention Acts, but added without sympathy that

> there is always a tendency among men to escape from laws which control their neighbours: and those that feared the change might occasion them a little expense, almost fell in love with the smoke, declaring that it hurt neither them nor their neighbours, nor their clothes nor their gardens.

Anti-smoke laws even made their way into literature: we find the term 'unparliamentary smoke' in reference to a large black plume from a factory in Mrs Gaskell's novel *North and South* (1855).[27]

The attention of the law was directed not only at factory chimneys. De la Beche and Playfair were in no doubt about the fact that the households of London contributed very materially to the pollution of the city's air, but there seemed to be no sensible way to legislate against domestic smoke. The average citizen was as worried as ever about smoky fireplaces and the damage that they did to interiors. There were a number of stoves and grates available to overcome these problems, and although they were reasonably successful in terms of operation, it is unlikely that they were popular enough to make any significant difference to the smoke burden of the London atmosphere.

Despite the wisdom of the principles of smoke abatement, even if only to lower the consumption of fuel, many people, including Charles Wye Williams and the editors of *Chambers Journal*, felt that the legislative measures were unlikely to succeed. It is true that extraordinary difficulties faced the implementation of the early legislation with the result that the Improvement Acts had, in general, been a dismal failure. Although there were successes, they were few and far between.

The new Acts were used to considerable advantage by the Reverend Septimus Hansard who took possession of the rectory at Bethnal Green in 1865. When he arrived he was told that not a blade of grass would grow in the sour soil. People laughed at the idea that he intended to grow flowers. Surrounding his 4 acres of open land were many houses and no less than eight factory chimneys emitting smoke at low level. He and many of his neighbours, who had long suffered the besooting effects of the smoke and the unhealthiness of the atmosphere, kept the police informed of every infringement of the Smoke Nuisance Acts. They realized that it was simply beyond the abilities of the police to notice all the breaches themselves, but that they were willing to act in response to complaints. The Reverend also found that the owners of the offending chimneys were, for the most part, ready to comply when due attention was drawn to them. Obviously it was in their own financial interest to abide by the law, if only to save fuel; it was presumably only apathy that had prevented them from meeting the sensible requirements of the Acts. The result of these efforts was, after many years, a rectory lawn fit enough to play croquet on, and asphodel, Greek acanthus and *Phlox drummondii* which were the envy of many London gardeners.[28]

Now that smoke abatement legislation existed, the police appeared to show no favouritism towards the factory owners[29] and those who persisted could use the laws to some effect. In addition to rising public awareness there was some interest on the part of factory owners to lower their own emissions. However it would be almost a century before the application of air pollution control legislation would become merely an administrative matter rather than the result of the extraordinary drive of small groups of people.

Notes

1. Pocock, D. C. D. (1979) 'The novelist's image of the North', *Trans. Inst. Brit. Geographers*, 4, 62–76, and classical comments from Defoe, D. (1724–7) *A Tour Thro' the Whole Island of Great Britain*, London; Cobbett, W. (1830) *Rural Rides . . . with Economical and Political Observations*, London.
2. A collection of American comments can be found in Allen, W. (1971). *Transatlantic Crossing*, Heinemann, London which includes selections from James Fenimore Cooper, Ralph Waldo Emerson, Frances Parkman and Bayard Taylor, all containing comments on the state of the air. Other references include Aderman, R. M., Kleinfield, H. L. and Banks, J. S. (eds) (1978) *Washington Irving's Letters*, vol. I (1802–23), Tawyne Publishers, Boston – letter of November 1805; and Hibbert, C. (1968) *Louis Simon: an American in Regency England*, Robert Maxwell, London.
3. Byron, Lord G. (1819) *Don Juan*, London.
4. In common use; for usage at the turn of the century see Laffin J. (1973) *Letters from the Front*, Dent, London.
5. Bevan, P. (1872) 'Our national coal cellar', *Gentleman's Magazine*, NS9, 268–78.
6. de Vries, A. (1974) *Dictionary of Symbols and Imagery*, North Holland Pub. Co., Amsterdam.
7. For example, *Cal. Pat. Rolls* 13 Ed. I m12; *Cal. Close Rolls*, 35 Ed. I m6d and m7d; subsequently *Cal. Pat. Rolls* 35 Ed. I m5d; *Cal. Close Rolls* 4 Ed. II m23d.
8. Nourse, T. (1700) *Campania Foelix*, London.
9. Wilde, O. (1891) *The Picture of Dorian Gray*, Ward Lock Ltd., London.
10. Tillman, D. A., Sarkanen. K. V. and Anderson, L. A. (eds) (1977) *Fuels and Energy from Renewable Resources*, Academic Press, New York; Slesser, M. and

Lewis, C. (1979) *Biological Energy Resources*, E & F. N. Spon, London; Barnaby, W. (1978) 'Sweden's sunny future', *Nature*, 273, 22 June.

11. Evelyn, J., *Diary*, 11 July 1656.

12. Owen, G. (1595) *A History of Pembrokeshire* (1595). The MS was printed in the *Cambrian Register* in 1796.

13. *A Trip to Scarborough*, III. iii. in Crompton Rhodes, R. (1962) *The Plays and Poems of Richard Brinsley Sheridan*, Russell and Russell Inc., N.Y.

14. *Gentleman's Magazine*, 24 (1754), 172.

15. Brown, S. C. (ed.) (1968) *The Collected Works of Count Rumford*, The Belknap Press of Harvard University Press, Cambridge, MA is a useful reprint of Rumford's writings which are now rather difficult to find. Instructions to build your own Rumford fireplace are given in Vivian, J. (1976) *Wood Heat*, Rodale Press, Emmaus, PA. There is a short discussion of his contribution in Pollock, W. F. (1881) 'Smoke abatement', *The Nineteenth Century*, 9 (March), 478–90. This may have sparked off an enquiry into the whereabouts of Rumford's stove which had been demonstrated by Faraday at the Royal Institute as late as the 1860s; see Anon., 'Smoke abatement', *Nature*, 293 (1882). An influential work on the subject of chimney design in the mid-nineteenth century was Arnott, N. (1855) *Smokeless Fireplaces and Chimney Valves*, Longman, London.

16. See Marsh, A. (1947) *Smoke*, Faber & Faber, London.

17. See B. White's preface to the 1772 edition of *Fumifugium*.

18. London Assize of Nuisance 1301–1431, *Lond. Rec. Soc.* (1973) in a case heard on 5 October 1377. However, most building regulations about chimneys concern not height but prevention of fires; see Articles of Wardmotes in Riley, H. T. (1861) *Liber Albus*. The Thomas Legg incident is quoted in the opening of Ashby, E. and Anderson, M. (1981) *The Politics of Clean Air*, Oxford University Press.

19. Frend, W. (1819) 'Is it possible to free the atmosphere of London from smoke . . .?' *The Pamphleteer*, 15.

20. Malcolm, C. V. (1976) 'Smokeless zones – the history of their development', *Clean Air*, 6 (23).

21. Hamilton, H. (1917) *Scientific Treatise on Smoke Abatement*, Sherratt & Hughes, Manchester.

22. Whytehead, W. K. (1851) *The City Smoke Prevention Act*, London.

23. Williams, C. W. (1841) *The Combustion of Coal and the Prevention of Smoke*, London.

24. The development of Victorian legislation outlined here comes mainly from the excellent paper by Ashby, E. and Anderson, M. (1976) 'Studies in the politics of environmental protection: the historical roots of the British Clean Air Act 1956.' *Interdisciplinary Science Reviews*, 1, 279–90.

25. Cutler, J. (c. 1864) *One Hundred and Fifty Six Letters Reporting on the Advantages and Disadvantages of a Grate*, privately printed.

26. *Chambers Journal* 19 (1853), 245.

27. Gaskell, E. C. (1855) *North and South*, London.

28. Pollock, W. F. (1881) 'Smoke Abatement', *The Nineteenth Century*, 9 (March), 478–90.

29. Carpenter, E. (1890) 'The smoke plague and its remedy', *Macmillan's Magazine*, 62, 204.

6

Smoke and the London fog

It seemed in the period that followed the passage of the Smoke Nuisance Abatement Acts that it would be only a matter of time before London's air pollution problems were solved. It was thought that, given sufficient vigilance on the part of the public and the authorities, the legislation should work. Engineers and technologists had already constructed a number of devices to lower the smoke emissions from chimneys. Many felt that the technology of smoke control was quite adequate and that it was merely a question of getting people to comply.[1] Common sense and good practice would solve all the problems in the end.

Despite these favourable prospects, it was apparent by the late 1870s that progress in air pollution abatement was not as great as had been hoped. It was true that increasing numbers of factory owners were concerned about the emissions from their factories and were making some efforts to lower them, even if only to achieve higher profits. Some industrialists were actively involved on Smoke Abatement Committees and it is likely that considerable reduction in the emissions from their factories was achieved after the passage of the early Acts. Following the Smoke Nuisance Abatement Act of 1853 there were smoke clauses in the Sanitary Acts of 1858 and 1866 and in the Public Health Act of 1875. In many cities, however, complications with local legislation meant that the laws on smoke abatement were often ignored. London was a little more fortunate. Here at least the laws were not ignored, but although the police were diligent in tracking down offenders, the technical issues involved meant that magistrates became reluctant to impose fines that were large enough to discourage pollution of the air. So by the 1880s it appeared that the enthusiasm for smoke abatement had lost some of its early drive.

As there was no air pollution monitoring network within London in the late nineteenth century, nobody really knew how much pollution was in the city's air. A few samples of rainfall had been taken and analysed by R. A. Smith, the first Alkali Inspector, in 1869 and 1870, but these isolated measurements could not indicate improvements or other long-term changes.[2] While the factory owners could point to the huge chimneys and show that there had been some decline in the emission of black smoke, there was a strong undercurrent of feeling that things had not really improved very much. True, it seemed that the black soot that dropped in flakes from the sky was not so noticeable where care had been taken to curb the worst emissions. Was it that the eye was easily fooled or were soot and smoke not the only important pollutants? One reason why many people thought pollution had actually got worse was that it seemed as if the climate of London was changing. The frequency and thickness of London fog had changed. Gloom, while indefinable in meteorological terms, had increased within the city.[3] These changes were intuitively linked to London's air pollution.

The history of London fog

There had always been fogs along the Thames, but somehow they had become particularly notable in the nineteenth century. At this time people began to sense that the fogs were related to air pollution and, as we saw in Chapter 2, high levels of pollution do aid the formation of fog. The fogs of the nineteenth century were thicker, more frequent and of a different colour from those of the past. It was difficult to know how long ago all these changes had started. Some of the earliest records of London fog come from the notebooks of the famous astronomer Thomas Harriot (1560–1621). He was using one of the earliest telescopes in England to observe the face of the sun. To cut down the brightness of the solar disc he had to observe it through mist or cloud, so his records contain numerous references to the clarity of the atmosphere in the first decade of the seventeenth century.[4] Fogs were by no means infrequent, but they were probably of quite natural origin. However, by the end of the seventeenth century it is no longer as easy to be so sure that the city was not partly responsible for some of the worst fogs. As we have seen, the nautical astrologer John Gadbury[5] noted some of the very thick persistent London fogs as 'Great Stinking Fogs' in his weather diary of the late seventeenth century, and H. R. Bentham, a German visitor of the 1680s, remarked on London's fogginess in his travelogue.[6] The high frequency of references to fog in London in the late seventeenth century suggests that it was rather more foggy than was to be expected. This may have been the result of a more stable pattern of atmospheric circulation which would have inhibited the dispersal of fog during this part of the Little Ice Age.

It became evident to a number of people interested in smoke abatement in Victorian England that long-term records of the fogginess of London would be very useful in determining changes in the quality of the city's air. Rollo Russell searched for records without success.[7] The meteorologists Mossman and Brodie were more persistent. Mossman was able to assemble

FIGURE 6.1 A fog on the Thames

a 200-year data set from old diaries and registers. He came to the conclusion
that there had been a startling increase in the fog frequency over that
period.[8] Brodie utilized the official records between 1870 and 1890 and
could detect an increase even over that relatively brief period[9] (see Table
6.1). The conclusions of Mossmann and Brodie met with considerable

TABLE 6.1 The increase in London fog, 1871–90, expressed as days with fog per year

1871–5	1876–80	1881–5	1886–90
51 ± 15	58 ± 15	62 ± 7	74 ± 11

Source: Brodie, F.J. (1892) 'The prevalence of fog in London during the twenty years 1871–1890', *Quart. J. Roy. Met. Soc.*, 18, 40–5.

resistance from meteorologists of the time who did not trust non-instrumental observations of fog, claiming that they were very subjective. The rejection of the idea that fogs were increasing in frequency also reflected a feeling among meteorologists of the time that, although climate fluctuated, it did not change in any systematic way.

The work of Mossman can be extended to give an indication of the frequency of London's fogs back well into the seventeenth century, through the use of astro-meteorological registers. It can also be brought up to date by using official meteorological records. Over the 300 years such an extended record covers, observation techniques, observer location and the geographical characteristics of London changed enormously. It is not simple to correct for these factors, so the raw data collected by the numerous observers must be interpreted with some care. Much of this data is summarized in Fig. 6.2a which shows the average number of foggy days per annum estimated from a large number of daily observations. The very earliest observations plotted in the figure suggest that fog frequency was rather higher then than at the start of the eighteenth century. This was the early period of frequent fogs already alluded to. Despite the possible urban origin of some of the 'Great Stinking Fogs', it is hard to apportion the blame for them. Pollution may have been responsible, but they may also have been frequent as a result of the stagnant circulation that persisted in the late seventeenth century.

Much more conclusive, in terms of our interest in urban climate, is the gradual increase in fog frequency that continued for more than a century between 1750 and 1890. During this period it is possible to be more certain that fogs were on the increase, because even in records from a single observer there are consistent rises in fog frequency. The most impressive record is that of William Cary, an instrument maker in the Strand, whose meteorological observations were published in the *Gentleman's Magazine* each month. They cover a period of some seventy years. Although the notion of a gradual increase in fogginess may well have been unpalatable to contemporary meteorological science, it appears that almost everybody in London was discussing the fogs.[10]

The fog frequency appears to have reached a peak in the 1890s, but the fogs were still frequent enough in the new century to warrant investigation. Just as the London Fog Inquiry (1902–4) got under way, the fogs began to disappear. This phenomenon has been discussed by Henry Bernstein in his article 'The mysterious disappearance of Edwardian London fog'.[11] The meteorologist F. J. Brodie, who had so carefully compiled records of the rising frequency of London fogs between 1879 and 1890, was quick to observe the decline in fog frequency in the early part of the twentieth century.[12] He attributed the change to the activities of the Coal Smoke Abatement Society which had been formed in 1899 and pressed for enforcement of the laws that required factories to consume their own smoke. It is also true that the emissions from domestic sources may have decreased through changes such as the wider use of the gas range. Although it would be satisfying to ascribe the improvement in the conditions to the efforts of the smoke abatement groups, it seems unlikely that these were the major causes of so marked a change.

The nineteenth century saw an enormous increase in the size of London as suburban transport networks enabled people to move about a larger city. As the metropolis expanded, the coal-burning activities of its inhabitants were gradually dispersed over a wider area. This would be expected to lead to a parallel decrease in fogginess. However, there may also have been a shift in the climate during the period which contributed to the sharp decline in fog.[13] The decline in fog frequency has continued throughout the present century and this has often been attributed to lower pollution levels as a result of the decreased emissions within London. Tighter controls on industry, changes in fuel type and the fact that coal was of declining importance as a domestic fuel must have helped lower the frequency of fog, but no one is really sure of the exact role that various factors played in the production of fog in Victorian London. Certainly by the time the Clean Air Act was passed in 1956 fogs had already decreased to a fraction of their late nineteenth-century frequency.

Regardless of the difficulties we currently find in understanding the origins of Victorian London fog there were fewer doubts at that time. Many influential people seem to have found it patently obvious that the increasing fog frequency was the result of increasing pollution of the air. This increase in pollution occurred in spite of the Smoke Nuisance Abatement (Metropolis) Act, which had been in force since the 1850s. The failure of the Act was not seen to be necessarily a result of weak enforcement or the pitifully small fines imposed by magistrates, because there was a strong feeling that factory emissions had in fact been considerably reduced. This was probably true because many factory owners had begun to 'burn their own smoke' in the interest of economy. However, some of the people who pressed for changes felt that the real source of pollution simply had not been identified.

The city as a source

When attempting to improve environmental quality there can be nothing quite so disastrous or disheartening as mis-identifying the pollutant or its source. Many of the writers at the end of the last century felt that the Smoke Nuisance Abatement (Metropolis) Act was too narrow in identifying a few specific industrial sources, and that the Act should be widened so that it covered all the sources. Up until the second half of the nineteenth century few Londoners had regarded the city as a whole as a source of pollution. The law had merely taken the most convenient path and placed restrictions upon those industries where control was feasible or those which had only a weak parliamentary lobby. The lack of appreciation of the city as an *areal* source of pollution stems from the earliest writings about air pollution. John Evelyn's concern is always directed at the large isolated sources of smoke, such as factory chimneys,[14] as he claimed that 'the *Culinary* fires . . . contribute little'. The interest in point sources persists with the notion of 'Black Catalogues' of polluters in the century that followed. Some country dwellers were more sensitive to the idea of the city as a source of pollution and claimed they could smell the city. London was called 'the big smoke' or sometimes simply 'the smoke'. For rather the same reasons Edinburgh was termed 'auld reekie', but that city also suffered from the stench of sewage and garbage.[15] Despite the fact that the cities earned these nicknames in the eighteenth century, it was only in the nineteenth century that the idea of the whole city as a source of pollution began to take on importance in air pollution science.

Early in the century the London meteorologist Luke Howard became interested in urban climatology. Howard is best remembered for the lasting contribution he made to meteorology by proposing the system of nomenclature for clouds that remains in use today, but his contributions to urban climatology are not so well recognized. Many of the notes he made in his meteorological register show that he was aware of the effects that the city as a whole had on the climate.[16] His studies of the temperature in and around London convinced him that the city had an 'artificial excess of heat' which amounted to more than 2 degrees Farenheit in the winter months. He deduced that this arose as a result both of fuel consumption in the city and of the increased absorption of radiation by the urban surfaces. The smoke and fogs that hung about the city are also the subject of many observations. For instance, on 10 January 1812, a windless day, London was plunged into darkness for several hours. The lamps in shops were lit and pedestrians had to take great care to avoid accidents. Howard's register entry ends with the comment: 'Were it not for the extreme mobility of our atmosphere, this volcano of a hundred thousand mouths would in winter, be scarcely habitable.'

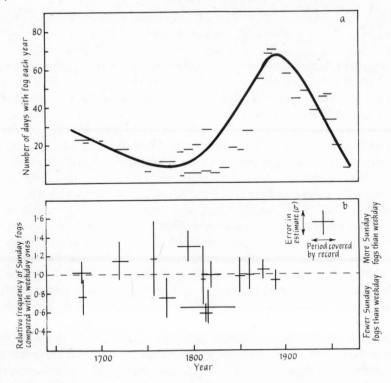

FIGURE 6.2 (a) Number of foggy days each year in London since the
seventeenth century; (b) relative frequency of Sunday fogs in London.
Values greater than 1 imply fogs were more frequent on Sundays; less
than 1, less frequent

Howard thought that all the London chimneys contributed to the
'fuliginous cloud' that so often hung over the city. At the end of the century
it had indeed become scarcely habitable during stagnant atmospheric
conditions. Other scientists were also aware of the magnitude of the
domestic source. Count Rumford claimed that many hundreds of tons of
wasted coal were suspended over the heads of Londoners at any given
moment.[17] No doubt he hoped that fear of its crashing down would
prompt Londoners to buy his model of smokeless stove. Howard, despite a
keen interest in anthropogenic effects on the atmosphere of London,
appears to have taken little interest in smoke abatement.

The failure of the legislation of the mid-nineteenth century caused a
number of other people such as Rollo Russell and Sir Francis Galton to
emphasize the areal nature of the source of London's pollution.[18] Both of
these men wrote about the very large and important contributions made by
the domestic consumption of coal. Russell strongly believed that the major
contributions were from the domestic sector, because there were more fogs

on Sundays and holidays than on working days. A reasonable argument it would seem, but quite the reverse of the one made by John Evelyn two centuries earlier. According to Evelyn, the industries were obviously the source of London's problem, because pollution almost vanished on Sundays. The frequency of fog on Sunday would seem a very useful index of the relative contributions of the domestic and industrial sector to the pollution over London, but when the available data is plotted out (Fig. 6.2(b)) there appears no strong evidence for either Evelyn's or Russell's claims. There may have been a slight decline in Sunday fog frequency at the start of the nineteenth century, but it is difficult to be certain even in this case.

Once again it is evident that the beliefs people held about London's weather were more important in influencing their actions than meteorological reality. As mentioned earlier, Russell tried to assemble a long record of fog frequencies but failed to find enough data. Therefore one must assume that his belief in the high frequency of Sunday fogs was based merely on a long memory. He used subjective evidence of this kind to argue for changes in the patterns of domestic energy consumption. While the meteorological observations Russell based on his memory may have been biased, he was probably right in assigning much of the smoke pollution to domestic sources. His pamphlet *London Fogs* was widely read in Victorian London and its impact has been compared with that of Rachel Carson's *Silent Spring* in our own time.[19]

The Fog and Smoke Committee

In the early 1880s the most active smoke abatement group in London was the Fog and Smoke Committee chaired by E. A. Hart, who ultimately became president of the National Smoke Abatement Institute. The committee always managed to get good publicity, have well-attended meetings and receive support from both nobility and politicians alike. From the start it seemed that although some tightening of the existing regulations might be possible with regard to the emissions from factories, there would be no way of extending these to cover domestic smoke. Enforcement, which was difficult enough with industrial sources, would become impossible when it came to the domestic hearth.

The Committee wisely pursued the problem of domestic emissions not through legislation, but by setting examples. They were aware, by this time, that the problem was not one of economics or technology, but rather of public attitude. While it might be possible to persuade people to use well-designed smokeless ranges for cooking, the cheery glow of the open fire in the living area was something that few were willing to part with. Some abatement enthusiasts maintained that the problem arose simply because servants didn't know how to make a good fire; perhaps this was the case with anthracitic coals and there was a suggestion that Welsh girls be sent down to the city to show the Londoners how to make a decent fire with the harder coal. The Committee's most successful public relations venture was the Smoke Abatement Exhibition which opened in South Kensington on 30 November 1881.

Although the 230 exhibits were divided into two main sections, one containing industrial equipment and the other domestic appliances, interest centred on the domestic apparatus. Each device was thoroughly tested by a group of experts and viewed by a critical public, who were quite sure that they wanted an 'open, pokeable, companionable fire'. The radiant source of heat provided by the open fire in the living room was regarded as very important. The heat radiating from the glowing fire warmed the walls and furnishings of the room rather than the air within it. This was thought to be much more pleasant than heating, which warmed the air by contact with hot surfaces. The use of the open fire was said to account for the general freshness of complexion of the inhabitants of the British Isles and the comparative infrequency of the need for spectacles among young people! Some people proposed theories that the 'Biovitric Rays' from the open fire were important for human fertility and in the present century Marie Stopes's name has been brought forward in support of the open hearth. With all the advantages attributed to the open fireplace is it any wonder that the simple stove never received wide acceptance?

The exhibition of 1881 was a failure to the extent that it did not encourage numerous London households to install smokeless grates. However it would have been far too ambitious to expect to see a widespread adoption of the new devices. The real importance of the exhibition was that once more the concept of cleaner air was given considerable publicity. Sadly, the years that followed the exhibition saw endless talk and little action. *The Times*[20] reminded its readers in 1883 that a century earlier Count Rumford had 'proved that the English grates and chimneys were fashioned as if deliberately to discharge most of the heat from the fuel into the air, and to retain most of the carbon as soot and dust', and still nothing had changed and the fogs continued.

London as a foggy city

London had gained its reputation as a foggy city from the German travellers of the late seventeenth century, and at first the visitors were disappointed when a fog restricted their view of the capital,[21] but by the nineteenth century many were even more disappointed if they were not confronted by 'London's Particular'. Somehow they felt that they had been cheated if they hadn't seen a fog, in much the same way as we would be a little disappointed if we arrived in Los Angeles and found no smog. In a letter dated October 1883 from James Russell Lowell the poet, who was US Minister to England, we find:

> To Miss Sedgwick   2, Radnor Place, Oct. 3 1888
> We are in the beginning of our foggy season, and today are having a yellow fog, and that always enlivens me, it has such a knack of transfiguring things. It flatters one's self-esteem, too, in a recondite way, promoting one for the moment to that exclusive class which can afford to wrap itself in a golden seclusion. It is very picturesque also. Even the cabs are rimmed with a halo, and people across the way have all that possibility of suggestion which piques the fancy so in the figures of fading frescoes. Even the gray, even the black fogs make a new and unexplored world not unpleasing to one who is getting palled with familiar landscapes.[22]

The season began in the late autumn and continued until winter had passed, but November was popularly considered to be the worst month, although once again the meteorologists disagreed. Certainly the novelists found the month to be cold and foggy and detective stories find it an essential background.[23] November was renowned not only for its fogs: Captain Frederick Marryat, author of *Children of the New Forest*, wrote that November was also the month of misanthropy and suicides.[24] It has been said that there was a French proverb which claimed that

> In October de Englishman shoot de pheasant
> In November he shoot himself[25]

A whimsical poem by Thomas Hood plays on the problems which beset the month:

> No sun, no moon . . .
> No leaves, no birds,-
> NOVEMBER

It is probable that the November fogs had particular impact because they were both thick and persistent (see Fig. 6.3). If only the days of very thick fog are considered then even in the records from the present century November is found to have the highest frequency. The November fogs still have a tendency to persist well into the day in the present century. It is evident that there is a difference between the human perception and meteorological observation. The meteorologist is more sensitive, as would be expected, and responds to less extreme conditions when recording fogs in his register. If the lighter mists noted by the experts are also counted as fogs then they become most frequent in December.[26]

Despite the disagreement over which month was worst, gloom covered the city in the winter months. Even the meteorologists were willing to

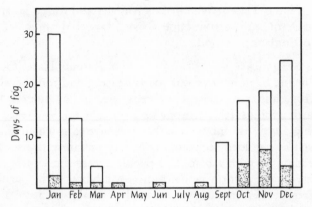

FIGURE 6.3 The seasonal distribution of fogs and thick fogs in London.
The shaded areas mark those fogs termed 'thick' in Luke Howard's
register

enter the word 'gloom' in their diaries with increasing frequency during the
course of the nineteenth century. The psychological and meteorological
gloom were no doubt interconnected as there are endless descriptions of
the dismal conditions that prevailed in the early part of the London
winter.[27] It was so dark that houses and shops had to be lit during the day.
This, of course, added to the cost of lighting. In fact interior lighting
during the day was not common in the late eighteenth century, but had
become rather frequent by Victorian times.

This meant that new terms such as 'day darkness' and 'high fog' began to
appear in the London vocabulary. The latter term, particularly, was used to
describe the occurrence of dark periods in the day, when no fog was
apparent at ground level.[28] At such times the sun was sometimes totally
obscured and, although it was very dark, it was still possible to see the
lights of buildings some miles away. As this phenomenon increased in
frequency, it caused problems to the early electric companies, whose
primitive switch-gear proved incapable of handling the surges of power
brought about by the sudden requirement for extra lighting. A contemporary record of the need for indoor lighting can be found in the
writings of a meteorologist, J. E. Clark, who, at the turn of the century,
recorded the times at which the light in his London office had to be turned
on each day.[29] The results of his investigations suggested that the need for
indoor lighting was unusually high in the morning. This was the time when
heavy fog was most frequent. It would appear from his records that
January was the gloomiest month in terms of the use of light, although the
record for December is lower than it might be because of the Christmas
holidays. Some time later instrumental records from the electricity
companies became available, and Fig. 6.6 shows the output from a
generating station on a day when conditions of high fog prevailed. Despite
the dramatic change in lighting requirement at the outset of the fog, the

FIGURE 6.4 Early morning fog and coffee

smoke concentrations, as recorded by early instruments, showed little change. A visual description of this particular incident makes reference to the relatively high visibility at ground level, so obscuration in this case seems to have occurred only at high levels.[30] This suggested that the smoke that caused the darkness was high up in the air.

Although such incidents are no longer very frequent, there was a notable

FIGURE 6.5 Boilers and coal supply for one of the early generating stations. These stations polluted the air, but also found it difficult to cope with the surges in electricity demand brought about by the rapid variation of 'London smog'

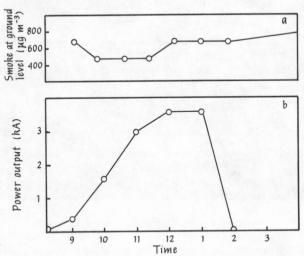

FIGURE 6.6 (a) Smoke concentration at ground level compared with (b) electricity generated. The rapid rise in electricity use occurs, on this occasion, at a time when there is little change in smoke at ground level. This is because a high-level smoke layer caused it to grow dark

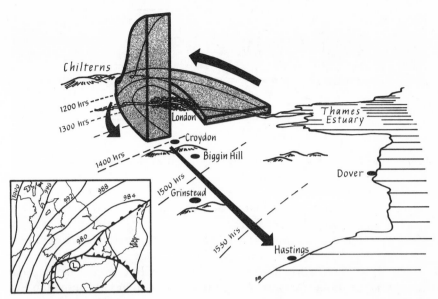

FIGURE 6.7 Diagram of air movement and weather map (inset) for dark day, 16 January 1955

period of day darkness on 16 January 1955. This particular event[31] occurred when extensive instrumentation was available. It is probably a reasonable model for earlier occurrences so it is worth examining in some detail. The weather pattern (Fig. 6.7) was characterized by a deep depression with rather weak pressure gradients near the centre (giving light winds) and very active fronts which gave rise to dense cloud. There was fog and a temperature inversion in the lower layers of the atmosphere. In the morning light, winds carried the smoke from London to the north-west. The smoke was unable to disperse because the inversion prevented vertical mixing. Estimates suggest that the original smoke layer was some 175 m thick. The movement of this air parcel containing smoke is illustrated in Fig. 6.7. The smoke reached the Chilterns just at the time when a cold front crossed them. It is possible that the smoke-laden air was lifted vertically in the vigorous convergence at the front. This would have stacked the smoky air in a vertical column more than a kilometre thick. Reports from aircraft suggest that the cloud layer was continuous and deep, stretching from 400–4000 m altitude.

About midday the smoke parcel began to pile up over the Chilterns. Shortly afterwards the wind direction reversed and gradually increased in strength. This carried the air, with the extremely dense pillar of smoke and cloud above it, back across London. The times at which the belt of extreme darkness reached various points in south-east England are shown in Fig. 6.7 and agree well with the expected movement of the smoky air under the

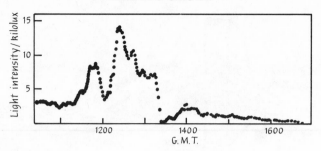

FIGURE 6.8 Light intensity through the dark day, 16 January 1955

prevailing winds. In London, records of both the light intensity and smoke levels are available. The smoke concentrations at ground level were not particularly great. The obscuration arose through the great vertical thickness of the smoke layer rather than its concentration at ground level (see Fig. 6.7). The light intensity on a sunny January day would have been about 36 kilolux, while illumination on a heavily overcast January day would fall to about 7 kilolux. At about 13.15 the light level shown in Fig. 6.8 dropped from 7 kilolux to less than 0.03 kilolux. Almost total darkness followed for six minutes. People who experienced this phenomenon said it seemed as if the world was coming to an end.

The effect of the fog

We have discussed the gloom that the fogs cast over the London winter and mentioned the psychological implications which this had. No doubt it affected tourism, but then maybe some people actually wanted to see the fogs. The increased need for lighting and the sudden changes in electrical load resulted in the initiation of considerable meteorological research by electric lighting companies. Additional cost of gas lighting for a single day of fog was put at £5000, but the total cost of a day's fog was thought nearer £20,000–£50,000 when disruptions to transport, accidents and extra cleaning required because of the sooty deposits left afterwards were included.

The effect of fog on transport had been a problem from the earliest times, and was even noted in records of seventeenth-century London life. The antiquarian Anthony à Wood records a great mist in London,[32] on 11 November 1667, when 'horses ran against each other, carts against carts, coaches against coaches, etc.' Evelyn tells of another similar event.[33] While thick fogs may have hampered travellers of Evelyn's time, they brought complete chaos to the transport system of Victorian London. 'Fog on the line' was a convenient excuse for sleepy apprentices arriving late at work, but there was more than just delay as a result of the fogs: difficulties in signalling meant an increase in railway accidents.

In an old issue of *Punch* there is a cartoon featuring two shadowy gentlemen; the caption reads, '*Befogged Pedestrian:* "Could you direct me to the river, please?" *Hatless and dripping stranger:* "Straight ahead. I've just come from it."' This is typical of *Punch* humour of the time, but a group of seven people did once walk into the Thames because of the fog.[34] In more tragic incidents in the fog that ran from 8 to 14 December 1873, no less than fifteen people were said to have drowned in the Northside docks. At Wapping two men walked into the river and drowned. Two workmen on their way home in St John's Wood walked into Regent's Canal and also perished.

There was increased mortality from disease during times of fog. Of course there is always an increase in mortality during the winter, but it was no longer possible to dismiss some of the increases as random fluctuations. In the week of fog in 1873 it appears that there were 700 more deaths than normally expected in London at that time of year. The public began to realize that there was a lot more to the fog nuisance than the transitory choking sensation every time they went onto the streets. Many people were dying in the smoke and fog, and many more were 'thrown so much out of health that they did not recover for some weeks'. The incident of 1873 was not the last by any means. The next two decades saw not only the increasing frequency of fogs, but the recurrence of these 'great fogs' that settled in over London for many days, when stagnant conditions allowed the pollutant concentrations to build up to high concentrations. These periods of continuous fog were followed by increased rates of mortality. Such episodes did not end with the Victorian era, as there have been a handful of bad fogs in this century. The worst of these fogs was 'The Great Smog' of 1952. In its wake there were some 4000 excess deaths. Important London fogs are summarized in Table 6.2.[35]

Naturally, animals were also affected by the fogs. In the fog of 1873 many of the cattle which were in London for exhibition at the Great Show at Islington are said to have died of suffocation. A large number of the beasts still living had to be put out of their misery. This incident seems to have left a particularly strong impression at the time; perhaps because, as one Irishman put it, the cattle had to be slaughtered not so much to save their lives, but for the value of their meat. We are told that sheep and pigs were less affected.[36]

Description of London fogs in literature

The high incidence of references to fog in the literature of London from the period of Dickens to that of T. S. Eliot is quite striking. Persistent fogginess provides a backdrop onto which we still project our vision of the life of Victorian London. Fog seems to be a prerequisite for detective stories set in London, and no period novel can any longer be written

TABLE 6.2 Major London smogs, 1873–1982

Year	Month	Duration (days)	Excess deaths	Maximum daily concentration SO_2 $\mu g/m^3$	Smoke $\mu g/m^3$
1873	Dec.	3	270–700		(a)
1880	Jan.	4	700–1100		
1882	Feb.				
1891	Dec.				(b)
1892	Dec.	3	~1000		
1948	Nov.	6	~300		
1952	Dec.	5	4000	3700	4460
1956	Jan.		480	2800	1700
1957	Dec.		300–800	2800	3000
1962	Dec.	4	340–700	4100	1900
1975	Dec.	3	(c)		500–600
1982	Nov.			560	

Notes:
(a) Smoke levels in the early fogs were 800 $\mu g/m^3$ or greater.
(b) The soot deposit during this fog was 9.4 g/m^2.
(c) Not statistically significant.

without reference to it. A reviewer of Robert Lee Hall's recent book *Exit Sherlock Holmes* recommends the book because it has the authentic fog and hansom cab flavour.[37] However, this image of fog-bound London may be stronger for us than it was for the Victorians. Can we imagine a Jack the Ripper story without its fog? While fogs are frequent in Sherlock Holmes stories, they are even more essential to the modern pastiches, if they are to capture the atmosphere of the period.[38]

These literary descriptions of London weather become very useful in the serious study of fog because there are so many gaps in contemporary scientific work on fog in the nineteenth century. It is true that the frequency of fogs can be obtained from the analysis of early meteorological records, but their distribution is more difficult to establish because of the paucity of meteorological stations. For instance, some meteorologists felt that the London fogs originated in the low-lying areas outside London and were then advected into the city by gentle winds. Non-meteorological sources[39] offer a contrary picture that suggests that fogs were usually centred on London. In Conan Doyle's *The Sign of Four*, the September evening proves to be a foggy one when Sherlock Holmes, Watson and Miss Morstan travel out to Upper Norwood, but by the time they pass Norwood the fog is far behind.

In the scientific literature there seems to be no satisfactory explanation for the colour of the fog. The earliest descriptions in diaries are vague, but as time went on yellow fogs became more frequently noted. They were also recorded several times, in the first decade of the nineteenth century, by the London meteorologist Thomas Forster.[40] A little later they were observed by his colleague Luke Howard,[41] but in general meteorologists didn't pay very much attention to the colour of the fog.

Although at the start of the nineteenth century Byron said that London had 'a dun coloured cupola', it wasn't until the 1840s that the thick yellow London fog really arrived. From then on the colour seems to have become ever more notable.[42] In a book published early in this century the writer claims that Londoners of the 1850s experienced the yellow fogs only rarely.[43] The coloured fogs would appear to have become not only more frequent, but also more strangely coloured, as may be seen in E. F. Benson's description.

> A sudden draught apparently had swept across the sky, and where before the thick black curtain had been opaquely stretched, there came sudden rents and illuminations. Swirls of orange-coloured vapour were momentarily mixed with the black, as if the celestial artist was trying the effects of some mixing of colours on his sky palate, and through these gigantic rents there suddenly appeared, like the spars of wrecked vessels, the chimneys of the houses opposite. Then the rents would be patched up again, and the dark chocolate-coloured pall swallowed up the momentary glimpse. But the commotion among the battling vapours grew ever more intense: blackness returned to one quarter, but in another all shades from deepest orange to the pale gray of dawn succeeded one another.[44]

It seems that these vividly coloured London fogs have almost vanished. Some argue that the fog of 1952 was yellow coloured, but the pea-soupers, named both for their thickness and colour, are in reality a thing of the past.[45]

The source of the colour is difficult to ascertain, but a number of possibilities spring to mind. It is conceivable that fine smoke particles in the atmosphere could absorb the blue wavelengths from the sunlight above the fog in such a way that fog at ground level was illuminated by a yellow light. This effect was noted as long ago as the Great Fire of London when the smoke in the air reddened the disc of the sun. The explanation gains some favour when Victorian writings are examined, but it cannot explain the yellow colour of the fog at night. However, it is possible that the fog at night was yellowed by the light of the gas lamps and the glare from shop windows.

It is also possible that the colour might have been the result of tarry compounds present in fog droplets. It is not known which of the tarry substances arising in low-temperature combustion of coal on the domestic hearth could give it the strange yellow colour, but after the fogs of 1891 Sir W. Thiselton-Dyer found tar to be deposited on the glasshouse at Kew at a loading of six tons per square mile.[46] In the Sherlock Holmes story, 'The Adventure of the Bruce-Partington Plans', we hear of a fog settling in over London in the November of 1895. On the fourth day of the fog Watson, through his intermediary and literary executor, Conan Doyle, wrote that: 'we saw the greasy heavy brown swirl still drifting past us and condensing into oily drops upon the window panes'.[47] This suggests that the droplets may have actually contained coloured substances. After the fogs of 9–12 December 1924 the roads were covered with a thin oily-looking film, according to the *Report of the Advisory Committee on Air Pollution*, and a later report gives just a little more detail. Here they were able to examine some of the fog droplets under a microscope and some of them were actually yellow and oily.[48]

The rapid changes in the intensity of the fogs are as evident in literature as they were in the records of the electricity companies. We read that

> From a sick dead yellow the colour changed to gray, and for a few moments the street seemed lit by a dawn of April; then across the pearly tints came a sunbeam, lighting them with sudden opalescence. Then the smoke from the house opposite, which had been ascending slowly, like a tired man climbing stairs, was plucked away by a breeze, and in two minutes the whole street was a blaze of primrose-coloured sunshine.[49]

In more sober literature, the concern was often with the psychological effect of the gloom and the sense of foreboding it cast over London in the foggy season. On his arrival in London in 1822, the diplomat Philipp von Neuman wrote in his diary for New Year, 'Everything here is empty; the fog and smoke obscure the atmosphere; there is a sad lugubrious air over everything, while Paris is all gay with plenty of Movement.'[50]

The role of fog in fiction

No doubt the fogs added a sense of mystery to the urban landscape and provided convenient cover for criminal activities, so it is understandable that fogs became an essential element of detective and mystery stories.

Fogs also occur as a portent of foreboding and gloomy prospects in Victorian novels. For instance, fogs are found at the beginning of *Bleak House*[51] and *A Little Princess*.[52] Here it is easy to see the fog as a metaphor for uncertainty about the future. In *Bleak House* the presence of smoke and fog is introduced again when Esther arrives in London and thinks that all the smoke is due to a fire. She had not realized that 'London's Particular' was always present. The notion of being 'fogged', that is confused, is an accepted metaphor that gained much currency in Victorian writing.[53] An extended use of fog as a metaphor for uncertainty occurs in E. F. Benson's *The Book of Months*, which opens:

JANUARY: Thick yellow fog, and in consequence electric light to dress by and breakfast by, was the opening day of the year. Never, to anyone who looks at this fact in the right spirit, did a year dawn more characteristically. The denseness, the utter inscrutability of the face of that which should be, was never better typified. We blindly grope on the threshold of the future feeling here for a bell handle, here for a knocker, while the door stood still shut.[54]

Despite the evils of the London fog, not everyone came out against it.[55] There was an ambiguity in the response to pollution that has already been noted in our discussion of the attitudes of Charles Dickens and Charles Lamb to air pollution in London. It seemed that many liked the brilliant, ever-changing spectacle of the fog. Others found that it added a sense of mystery and charm to the London scenery and buildings. M. H. Dziewicki even wrote an essay 'In praise of London fog'.[56]

Fog may also be an element of an apocalyptic mood in literature and art that is to be found at the end of the nineteenth century.[57] While the apocalypses we imagine today are global, the Victorian examples were sometimes rather regional or even urban. Richard Jefferies's novel *After London* (1885) and Robert Barr's story 'The Doom of London' (1892) concern the destruction of the city.[58] Poets of Victorian London frequently wrote on similarly destructive themes[59] and some of the titles, such as *The Doom of the City: A Fantasia*, convey the idea.

Why did some Victorian writers have these apocalyptic visions of London's future? Many had an intense dislike of cities, but it is also true that they were influenced by the pollution and the climatic change they saw in nineteenth-century London. Richard Jefferies' notebooks reveal that he saw London as disease-ridden, and in a short piece of writing called 'The Great Snow', probably written before 1875, he has London destroyed in a massive snowstorm. This is in line with the impression of particularly harsh winters we found in Sherlock Holmes stories. When Felix, the hero of *After London*, visits the ruins, he only narrowly escapes the poisoned air that hangs over the remains of the metropolis.

The most pertinent example of a story which brings home the link between air pollution and the destruction of London is Barr's 'The Doom of London'. Barr is better remembered for his novels and detective stories these days, but 'The Doom of London', written after a bad series of fogs, was a chilling and prophetic vision. With Barr's story the link between pollution and destruction of the city is more direct. Almost the entire population of London dies of asphyxiation in a fog that envelops the metropolis. Barr admitted that his vision of the future arose out of the frequent London fogs and he republished the story in the *Idler* of 1904 after some particularly intense fogs.

'The Doom of London' is set in the future, probably about 1940. It purports to be written by an old man, who knew the London of the 1890s. He reflects on the most catastrophic of all London fogs and describes how the early part of Victorian November had been very fine, but one Friday a fog settled in. It wasn't a very bad fog by London standards, but it persisted, and day by day grew thicker with the enormous amount of coal smoke that poured into it. The air was peculiarly still. As always in a fog, the death rate increased, but the increase was no more so than usual until the sixth day. On the seventh morning of the fog, the newspapers were full of startling statistics, but their significance was not realized. Everywhere people started to die, but the storyteller was fortunate enough to have a respirator. The oxygen was being used up, so gradually the gas lights were extinguished and only the electric lights remained. In Cannon Street he 'ran against a bus, spectral in the fog, with dead horses lying in front, and their reins dangling from the nervous hand of a dead driver. The ghost-like passengers, equally silent, sat bolt upright, or hung over the edge-boards in attitudes horrible grotesque.'

He managed to escape on a train, out of control, that left from Cannon Street station, down tunnels that still contained a little fresh air brought in from the country. Although the train was quite packed with people only two survived. Soon after his escape from London, a westerly gale cleared the air and another 167 persons were rescued from a fearful heap of dead on the platforms of Cannon Street station. In all, few Londoners survived. Looking back on the destruction of London, many of the 1940s academics in the story saw it as an unmixed blessing. Like the destruction of Pompeii, the destruction of London was viewed as the destruction of vice and money-gathering.

Robert Barr thought that the stagnant air would be depleted of its oxygen to the last atom. We know that this could not happen, because combustion processes and respiration would cease long before all the oxygen had been consumed. More important, for human life at least, the build-up of toxic gases and particulate materials would act much more rapidly than asphyxiation. The story, despite both literary and scientific flaws, is fascinating because it characterizes some of the feelings of London's impending doom that appear in the nineteenth century.

Conan Doyle also attempted an apocalyptic work, *The Poison Belt*.[60] In this short story the end of human life on Earth almost comes about as the Earth passes through a belt of anaesthetic gas. *The Purple Cloud* by M. P. Shiel is the fully apocalyptic development of this theme.[61] Here the entire animal population of the Earth, bar the storyteller, is annihilated through volcanic emanations of cyanides. In rather the same way as *After London* and 'The Doom of London', the book is set some time in the future. There is no direct link to the pollution of the time in this book, but it is steeped with observations of recent geophysical events. The vivid sunsets that were seen in England after the eruption of Krakatoa in the 1880s may be evident in the book: vivid sunsets occur in the story after the eruption responsible for the release of the annihilating cyanide gas. The book continues with the Earth's sole survivor undertaking a campaign of city-burning in which he destroys, among others, London, Paris, Calcutta, Istanbul, San Francisco and Peking. The destruction of cities and the vice and disease that they represent is an important feature of many of these apocalyptic writings. These authors show an intense dislike of the cities they destroy within their fiction.

Artistic impact

This apocalyptic vision also found expression in a number of works of art of the nineteenth century. Views of the destruction of great classical cities were a favoured theme. With respect to London one of the most imaginative of these visions is the engraving by Gustave Doré, The New Zealander, set in the future 'when some traveller from New Zealand, in the midst of vast solitude, take his stand on a broken arch of London Bridge to sketch the ruins of St Paul's'.[62]

However, some Victorians were concerned that the fogs were having a negative effect on the art of the nation. Not only were the fogs damaging the paintings, sculpture, leather furniture and book bindings in a material way, but they were preventing the artists from working. Sunshine and the blue skies and hard shadows it brought were less frequent, and often it is said that the artists couldn't even see their subjects.[63] It is true that blue skies had become less frequent in paintings by the nineteenth century.[64] The predominance of blue skies in medieval paintings may have been due to the iconographic approach of the artists, in which the sky merely acted as a blue backdrop. Later this may have reflected a southern influence that led to a high frequency of Italian skies. In the eighteenth century real views of nature were not regarded as acceptable by painters such as Gainsborough, who thought the subject must come from the artist's own mind. Gradually, however, landscape painting, once a lowly form of draughting, became more acceptable, and painters began to question the relevance of Mediterranean light in an English landscape. Constable, strongly influenced by the flat East Anglian landscapes dominated by skies, made a careful study of the English sky and initiated an era that demanded more realism.[65] Small wonder that the frequency of blue skies in paintings began to decline. More relevant to an interest in air pollution is the increasing frequency of yellow winter scenes in the foggy cities; here the artists may well be responding to a changing urban meteorology. Some felt a little overwhelmed by all that blue. Giuseppe Mazzini (1805–72), the Italian patriot, said on his return to Italy that he grew tired of the eternal blue of the Mediterranean skies and wished for a little London fog.

FIGURE 6.9 Doré's apocalyptic vision of a future New Zealander sketching the ruins of London

The writers of the nineteenth century, who claimed that painting was being hampered by the dense atmosphere, failed to realize the enormous changes that art was undergoing. Painting had much changed under the influence of Turner and Whistler, so that absolute clarity was no longer essential. Great artists were seeing new opportunities in the polluted urban environment. Turner was obviously fascinated by steam and fog, and the Impressionists no less so. Monet visited the capital in the early part of his career and did not seem to be bothered by the fog at all: his painting *The Thames below Westminster*, which hangs in the National Gallery, is a good example. Meteorologists interested in art have noticed a gradual decrease in visibility in European paintings. Monet deliberately chose to visit London in the winter in order to paint his Thames series.[66] Other Impressionists, such as Pissarro, included the word 'fog' in the titles of their paintings. Fogs and air pollution did not stifle artistic development. The Chinese artist Chiang Yee, who was active in London of the 1930s, wrote of the fog in his book *The Silent Traveller in London* and found the fogs an inspiration and an aid to creating perspective in his oriental style.[67]

Notes

1. *Chambers Journal*, 19 (1853), 245–8; see also Whytehead, W. K. (1851) *The City Smoke Prevention Acts*, London; Williams, C. W. (1856) *Prize Essay*, Weale, London.
2. Smith, R. A. (1872) *Air and Rain*, London.
3. *Gentleman's Magazine*: see Carey's and later Gould's monthly meteorological tables.
4. Shirley, J. W. (ed.) (1974) *Thomas Harriot*, Oxford University Press.
5. Gadbury, J. (1691) *Nauticum Astrologicum*, London.
6. Bentham, J. R. in Robson-Scott, W. D. (1953) *German Travellers in England*, Basil Blackwell, Oxford.
7. This unsuccessful search is mentioned in a discussion that followed the presentation of Brodie's paper to the Royal Meteorological Society in 1891; see *Quart. J. Roy. Met. Soc.* 18 (1892), 44.
8. Mossman, R. C. (1897) 'The non-instrumental meteorology of London 1713–1896', *Quart. J. Roy. Met. Soc.*, 23, 287–98.
9. Brodie, F. J. (1892) 'The prevalence of fog in London during the twenty years 1871–1890', *Quart. J. Roy. Met. Soc.* 18, 40–5.
10. While there are numerous references to fog in the nineteenth century, the following are of particular interest. Howard, E. (1893) *The Eliot Papers No. 1*, John Bellows, Glasgow. Howard (a relation of Luke Howard) was convinced that fogs had much increased since the eighteenth century. Similarly Beale, S. S. (1908) *Recollections of a Spinster Aunt*, Heinemann, London. Hartwig, G. (1877) *The Aerial World*, Longman, Green, London claims that the fogs have a 'proverbial celebrity'.
11. Bernstein, H. T. (1975) 'The mysterious disappearance of Edwardian London fog', *The London Journal*, 1, 189–206.
12. Brodie, F. J. (1905) 'Decrease in London fog in recent years', *Quart. J. Roy. Met. Soc.*, 31, 15–28.
13. Lamb, H. H. (1977) *Climate: Present, Past and Future*, Methuen, London has much discussion on the Little Ice Age.

14. Evelyn, J. (1661) *Fumifugium*.
15. 'Reek' has definite associations with smoke, but we know of the odiferous nature of Edinburgh from many eighteenth-century sources; see the entry of 12 August 1733 in Hill, G. H. (ed.) and Powell, E. F. (reviser) (1950) *Boswell's Life of Johnson*, vol. V, *The Tour of the Hebrides*, Clarendon Press, Oxford.
16. Howard, L. (1833) *The Climate of London*, London.
17. *Chambers Journal* (1854), 106. This, like many of the estimates of trace components made in preceding centuries, is much too high. If we took London as measuring 5 km × 5 km on a square in Rumford's day (*c*. 1800), with an atmospheric mixing layer 150 m deep, then 100 tons of suspension would result in a mean smoke concentration in excess of 25,000 $\mu g/m^3$ Measurements made by W. J. Russell in the 1880s were 124, 324 and 862 $\mu g/m^3$ – the last value obtained during a fog. Russell's measurements seem reasonable. However it is interesting to note that even J. S. Owens, who did so much to set up the early monitoring network in the British Isles, was prone to exaggeration. He said that there were 200–250 tons of soot in the London air (*J. Roy. Soc. Arts*, 73 (1925), (434–53), which would imply concentrations far higher than those he was measuring.
18. Russell, R. (1889) *Smoke in Relation to Fogs in London*, National Smoke Abatement Institute, London (1889); Galton, D. (1880) *Preventible Causes of Impurity in London Air*, Sanitary Inst. of Great Britain; Russell, R. (1880) *London Fogs*, E. Stanford, London.
19. Ashby, E. and Anderson, M. (1977) 'Studies in the politics of environmental protection: the roots of the Clean Air Act, 1956: II. The appeal to public opinion over domestic smoke, 1880–1892', *Interdisciplinary Science Reviews*, 2, 9–26. This is a complete guide to efforts over the period.
20. *The Times*, 18 July 1883.
21. Pülker-Musmau, Prince, (1826–8) *A Regency Visitor*.
22. Allen, W. (1971) *Transatlantic Crossing*, Heinemann, London.
23. Of course classic November fogs occur in Sherlock Holmes stories, but we find important ones elsewhere; e.g. Perkins, C. L. (1894). 'The Redhill Sisterhood', in *The Experiences of Loveday Brooke, Lady Detective*, Hutchinson, London, which opens, 'It was a dreary November morning; every gas jet in the Lynch Court Office was alight, a yellow curtain of outside fog draped its windows', and the notion can be found persisting in literature to the present, e.g. Plath, S. (1965) *Ariel*, Faber & Faber, London, where the poem 'Letter in November' contains the phrase 'in thick grey death soup', written in the foggy period of the early 1960s.
24. Marryat, Capt. F. (1832) *Newton Forster; or the Merchant Service*, London.
25. Massingham, H. and Massingham, P. (1950) *The London Anthology*, Phoenix House, London.
26. Chandler, T. J. (1965) *The Climate of London*, Hutchinson, London.
27. Chancellor, E. B. (1928) *The Diary of Philipp von Newman*, Philip Allen, entry for 1 January 1822; and Luard, C. G. (1926) *The Journal of Clarissa Trant*, Bodley Head, London, entry for late 1819. Both these entries contrast the gloom of the London winter with the brightness of continental cities.
28. *The Advisory Committee on Atmospheric Pollution*, 10th Report (1925).
29. Clarke, J. B. (1901) 'Day darkness', *Met. Mag.*, 36, 194.

30. *The Advisory Committee on Atmospheric Pollution*, 10th Report (1925), 32.

31. Helliwell, N. C. and Blackwell, M. J. (1955) 'Daytime darkness over London, Jan. 16 1955', *Met. Mag.*, 84, 342.

32. Clark, A. (1892) *The Life and Times of Anthony à Wood, Antiquary of Oxford, 1632–1695*, Oxford Historical Society, vol. II, 121, an entry for 11 November 1677.

33. Evelyn, J., *Diary*, 25 November 1699.

34. Hartwig, G. (1879) *The Aerial World*, Longman, Green, London, and also Chiang Yee, (1938) *The Silent Traveller in London*, Country Life Books, London, which contains a more affectionate description.

35. Bach, W. (1972) *Atmospheric Pollution*, McGraw-Hill, New York; Ball, D. J. and Hume, R. (1977) 'Vehicular and domestic emissions of dark smoke', *Atmospheric Environment*, 11, 1065–73; Ball, D. J. and Schwar, M. J. R. (1983) *Thirty Years On*, GLC, London; Burgess, S. G. and Shaddick, C. W. (1959) 'Bronchitis and air pollution', *J. Roy. Soc. Health*, 79, 10–24; Gore, A. T. and Shaddick, C. W. (1958). 'Atmospheric pollution and mortality in the county of London', *J. Brit. Preventive Medicine Soc.*, 12, 104–13; Heinmann, H. (1961) 'Effects of air pollution on human health', *Air Pollution*, WMO, Geneva; Logan, W. P. D. (1949) 'Fog and mortality', *The Lancet*, 78; Logan W. P. D. (1953) 'Mortality in the London fog incident, 1952', *The Lancet*, 336–8; Read, B. (1970) *Healthy Cities: A Study of Urban Hygiene*, Blackie, Glasgow; Russell, W. T. (1924) *The Lancet*, 335–9; Russell, W. T. (1926) 'The relative influence of fog and low temperature on mortality from respiratory disease', *The Lancet*, 1128–30; Scott, J. A. (1963) 'The London fog of December 1962', *The Medical Officer*, 109, 250–2; Warren Spring Laboratories (1967) *Atmospheric Pollution 1958–1966*, 32nd Report, HMSO.

36. Hartwig, G. (1877) *The Aerial World*, Longman, Green, London; Meetham, A. R. (1964) *Atmospheric Pollution*, Pergamon Press, Oxford, 231, suggests that sheep and pig pens were not frequently cleaned and the ammonia from decaying urine and faeces neutralized the effects of the acid fog droplets.

37. See the cover of Hall, R. L. (1979) *Exit Sherlock Holmes*, Sphere Books, London.

38. While the modern reader may be more likely to notice the fogs in Sherlock Holmes stories, there is also a surprisingly high frequency of violent storms and harsh winters. Thesing, W. B. (1982) *The London Muse*, University of Georgia Press, Athens, GA notes 'the thematic use of natural disturbances such as storms and fogs' in the poems of Laurence Binyon's *London Visions* (1896 and 1899).

39. In addition to the *Sign of Four*, such a situation is also noted in Perkins, C. L. (1894) 'The Redhill Sisterhood' in *The Experiences of Loveday Brooke, Lady Detective*, Hutchinson, London, but perhaps the most conclusive evidence comes from the daily entries in Ponsonby, A., 'Meteorological Register 1884–93' (MS in my personal collection) which records the frequency of fog in London as three times that in Ascot.

40. Forster, T., 'Meteorological Journal kept at Clapton in Hackney' (see *The Gentleman's Magazine* for 1811).

41. Howard, L. (1833) *The Climate of London*, London.

42. There is a quotation 'No vapour'd foggy gloom imbrowns the sky' in Dodsley, R. (1782) *A Collection of Poems*, VI, London, but this is not typical of

the period. Vivid yellow fogs do not become frequent until the 1840s: e.g. Taylor, B. (1846) *Views Afoot*, London.

43. Beale, S. S. (1908) *Recollections of a Spinster Aunt*, Heinemann, London.
44. Benson, E. F. (1905) *Image in the Sand*, Heinemann, London. This may seem to be pretty wild writing, but similar descriptions of spectacular fogs are described by more restrained writers, e.g. Stevenson, R. L. (1886) *Dr. Jeckyll and Mr. Hyde*.
45. Bonacina, L. C. W. (1950) 'London fogs – then and now', *Weather*, 5, 91; also *Weather*, 15 (1960), 127.
46. Shaw, Sir N., 'The treatment of smoke: a sanitary parallel', *Nature*, 66, 667–70.
47. It is also worth remembering the striking image in T. S. Eliot's *The Love Song of J. Alfred Prufrock*, 'The yellow fog that rubs its back upon the window panes'.
48. *Advisory Committee Report on Atmospheric Pollution*, 9th Report, 46–59, and the suggestion of tar in droplets is also found in Lewes, V. B. (1910) 'Smoke and its prevention', *Nature*, 85, 290–5; Walter Scott (1892) says that the carbonaceous and tarry compounds make the fog yellow and dirty.
49. Benson, E. F. (1905) *Image in the Sand*, Heinemann, London.
50. Chancellor, E. B. (1928) *The Diary of Philipp von Neuman*, Philip Allen, London.
51. Dickens, C. (1852–3) *Bleak House*, Bradbury & Evans, London, published in parts.
52. Burnett, F. E. H. (1905) *A Little Princess*, Warne, London.
53. Chiang Yee (1938) *The Silent Traveller in London*, Country Life Books, London, Victorian in morality, even if not by date!
54. Benson, E. F. (1903) *The Book of Months*, Heinemann, London.
55. Duncan, S. J. (1891) *An American Girl in London*, Chatto & Windus, London; Cook, E. T. (1903) *Highways and Byways in London*, Macmillan, London. It is argued that smoke 'which spoils everything else, beautifies London by disguising its ugliness'; see Ewart, W. (1902) 'Report on the counties of London and Middlesex', in *Report of a Committee of the Royal Medical and Chirurgical Society of London*, Macmillan, London.
56. Dziewicki, M. H. (1902) 'In praise of London fog', in Singleton, E. (ed.), *London – as Seen and Described by Famous Writers*, Dodd Mead, New York.
57. Kermode, F. (1967) *The Sense of an Ending: Studies in the Theory of Fiction*, Oxford University Press.
58. Richard Jefferies (1885) *After London*, London; Robert Barr (1892) 'The Doom of London', in *The Idler*, 397–409. The latter story was reprinted a little later in *The Idler*, 26 (1904), 540, and a modern reprint is available in Lodge, J. P. (ed.) (1970) *The Smoake of London: Two Prophecies*, Maxwell Reprint Co. There are modern counterparts to these regional Victorian apocalypses. The apocalyptic fiction of J. G. Ballard features both global and more restricted types of apocalypse.
59. E.g. Noel, R. (1872) 'The Red Flag'. The poem, influenced by the Paris commune, ends with a cataclysmic vision of the destruction of the city. Another poem has wonderful images of the pollution within the city, which, as is frequent in Victorian literature, is compared to Babylon: see Noel, R. (1872) *A Lay of Civilisation or London*.
60. Doyle, A. C. (1913) *The Poison Belt*, Hodder & Stoughton, London. M. Paul Souday of *Le Temps* claimed, rather unfairly, that the story had been copied

from Rosny's *La Force mystérieuse*, which describes destruction in terms of the alteration of the properties of light.

61. Shiel, M. P. (1901) *The Purple Cloud*, Chatto & Windus, London. Shiel was a 'recklessly imaginative Edwardian romancer' it is argued in Hunter, J. (1982) *Edwardian Fiction*, Harvard University Press, Cambridge, MA, but despite this the series of his novels being republished by Gollancz in 1929 came to an abrupt end.

62. Macaulay, Lord T. B. (1840) 'Review of Leopold Von Ranke's "The Ecclesiastical and Political History of the Popes of Rome in the Sixteenth and Seventeenth Centuries", translated by Sarah Austin', *Edinburgh Review*, (Oct.), 62, the quotation here no doubt inspired the engraving.

63. Russell, R. (1889) *Smoke in Relation to Fogs in London*, National Smoke Abatement Institute, notes the expense and loss of time by artists because of the smoke-fogs.

64. Neuberger, H. (1970) 'Climate in art', *Weather*, 25, 46; J. S. Owens (1925) *J. Roy. Soc. Arts*, 73, 450 suggested that the smoke haze may have added atmosphere.

65. An excellent exhibition on the subject of painting and meteorology, 'The Cloud Watchers', was held in The Herbert Gallery and Museum in Coventry in 1975 and the catalogue of the same title is a valuable treatment of the subject.

66. Seiberling, G. (1981) *Monet's Series*, Gardland Publishing, New York and London.

67. Chiang Yee (1938) *The Silent Traveller in London*, Country Life Books, London.

7

Monitoring the changes in atmospheric composition

As emphasized in the previous chapters, there was no air pollution monitoring network in Victorian London. Trends in the atmospheric contamination could only be guessed at from the frequency of related phenomena such as fogs. With these it appears that the interpretation could depend very much on what the observer wished to prove. So in the late nineteenth century the notion of air quality remained based on perception, much as it was in John Evelyn's time. Consequently, in addition to the possibility of bias in the observations, attention focused on the more perceptible pollutants such as smoke. Sulphur dioxide, one of the most damaging components of air in coal-burning cities, tended to be overlooked.

This meant that the approach to air pollution in Victorian London not only mis-identified the sources of pollution, that is by concentrating on a few large factories rather than the city as a whole. It also meant that they very nearly identified the wrong pollutant by directing all their attention to the visible smoke, forgetting the invisible sulphur dioxide.[1] One must sympathize with their problem. They had enough difficulty in pressing for the reduction of smoke emission. It is easy to imagine how little success they thought they would have had in obtaining reductions in the emission of an invisible and, for all practical purposes, undetectable pollutant in the air of Victorian London. Even if the Fog and Smoke Committee had succeeded in their attempts to have a smokeless grate in every London home, there would have been little change in the levels of sulphur dioxide emissions, because this pollutant gas is released even when great care is taken to burn the smoke.

Fortunately some progress had already been made in legislating against invisible gaseous pollutants, in those cases where the effects were catastrophic. The emission of chemicals into the atmosphere had been the concern of a group of people somewhat removed from those who fought for the abatement of smoke. By the mid-nineteenth century large tracts of countryside about St Helens, Newcastle and Glasgow had been destroyed. Contemporary descriptions said it was as if the land had been swept by deadly blights till it was as barren as the shores of the Dead Sea. The problem arose from the emission of hydrochloric acid by the alkali industry. A Select Committee under the chairmanship of Lord Derby sat through the summer of 1862 to see what could be done to ameliorate the damage that was being caused.[2] The alkali industry was concerned with the production of salt-cake or sodium sulphate, much needed in glass manufacture, and the production of the alkalis such as sodium carbonate and sodium hydroxide. The method used had been developed by Leblanc and was based on the reaction of common salt with sulphuric acid according to the equation:

$$2NaCl + H_2SO_4 \rightarrow 2HCl + Na_2SO_4$$

The hydrogen chloride produced as a by-product was of little interest in the early days of the industry and was consequently discharged into the air. As the industry grew, the amounts of waste gas released became enormous, and the environmental damage correspondingly so.

The solution to the problem was simple. It merely required the fumes from the manufacturing process to be washed, then the acid vapours would be dissolved in water. The process was so inexpensive that a number of manufacturers had already adopted it to improve their image with neighbours, but there was no incentive to maintain the equipment properly. Because such a simple and effective control technology was available to the manufacturer, Lord Derby's Select Committee was actually able to contemplate legislation. It was a radical notion for the state to interfere with industrial processes and many felt that such actions would have a damaging effect on national prosperity. The annual output of alkali works was valued at some £2,500,000 and they employed 19,000 people, paying a total wage of £870,000. In August 1862 the Select Committee reported its findings to Parliament and the Alkali Acts became law in the following year.

The rapid passage of the bill through Parliament is in marked contrast to the problems encountered with the Smoke Nuisance Abatement (Metropolis) Act. The bill did not prove a very controversial one, partly because the alkali manufacturers admitted that the industry was responsible for considerable environmental damage and were already tired of being the subject of numerous complaints. Prior to the Alkali Acts they were sued for nuisance in actions that were difficult and frustrating. Their reason for refusing to control the emissions from their plants was merely one of competitive effectiveness. This remains a key issue in industrial emission control to this day. If a factory is obliged to control its emissions, then its costs are thereby increased, making it less effective in competition with similar factories that have not adopted the control measures. Thus it is only when there is a law to compel all manufacturers to curb emissions that equality is maintained. Another reason why the bill had a quiet passage through Parliament was that the modified bill removed dictatorial powers from the proposed Alkali Inspector and made it necessary to bring actions against the manufacturers through the county courts.

As Lord Ashby pointed out in the fourth W.E.S. Turner Memorial Lecture, the alkali legislation had three important features:

  (i) it challenged the social norm, suggesting that the destruction of vegetation or other amenities was not the inevitable price of industrial progress;
 (ii) it restricted only the emission of hydrochloric acid, where a control technology was readily available; and
(iii) it controlled the nuisance by requiring emissions to be cut by 95 per cent rather than prohibiting them.

The requirements of the legislation were hardly impossible to meet. It was such reasonableness or even laxity that enabled the law to operate. However it should be pointed out that some regard this as the very beginning of the weakness of British environmental legislation. True, it may seem undesirable that the offenders got off rather lightly, but the point of environmental legislation is not to promote litigation: it is to improve the quality of the environment, and that was the purpose of the Alkali Acts. An important feature of the Acts was the fact that the Inspectorate was attached to central rather than local government. This meant that uniform standards would be applied over the whole country and the inspectors would not be members of a local authority subject to pressure from local industrialists.

The technology to be adopted in emission control was not defined. Sir Lyon Playfair felt that any definite plan here 'might stop invention'. It was left to the manufacturers to arrive at satisfactory processes to remove the hydrochloric acid. The simplest arrangement was to pass the gas through large bottles containing water. The hydrochloric acid so formed was not very strong. More valuable, concentrated hydrochloric acid could be made by passing the hydrogen chloride through towers filled with coke or specially manufactured plates to increase the contact area.[3] Water trickled down over these surfaces and absorbed the gas.

Angus Smith

The country's first Alkali Inspector was R. A. Smith. He, like many of his successors, tried to be not too disagreeable in order to obtain the co-operation of the manufacturers. This 'unholy alliance', as it is often termed today, led to a marked reduction in the emissions of hydrochloric acid from the alkali plants. Despite criticism for being 'too soft', Smith was an

FIGURE 7.1 Angus Smith, the first Alkali Inspector

extremely astute man and saw far beyond the narrow confines of his particular brief, which was simply to reduce the emissions of hydrochloric acid from alkali works to 5 per cent of the original value. He thought that all noxious fumes were of public concern. There was also the problem that no matter how much damage was caused by the lowered emissions of hydrochloric acid, the manufacturer had fulfilled his statutory obligations merely by reducing the escape of acid by 95 per cent. Hence the Alkali Acts actually permitted the factories to continue to pollute. It is this early lesson that explains much of the subsequent reluctance to set legally permissible levels that so typifies much subsequent environmental legislation in the United Kingdom.

While the Acts were successful in reducing the emissions of hydrochloric acid by more than 95 per cent (in fact it was nearer 99 per cent), the improvement in quality of the local environment was not as marked as had been hoped. It was observed that roses did not grow where once hawthorn had struggled. There were other pollutants also. The first Alkali Inspector often encountered large emissions of sulphuric acid from alkali works. In some cases this caused more damage than the hydrochloric acid specified in the Acts, yet nothing could be done to curb these emissions. The need for great flexibility in the laws was self-evident. Smith's successor thought that manufacturers should be required to use the *best practicable means* to lower pollutant emissions, as this 'would prove an elastic band ever tightening as chemical science advanced and placed greater facilities in the hands of the manufacturer'.[4]

Prior to any modification of the original Alkali Acts, R. A. Smith was

TABLE 7.1 Oxygen and carbon dioxide at various locations as determined by
R. A. Smith

Oxygen:	%
Remote locations	20.990
Suburb Manchester in wet weather	20.970
Manchester in fog and frost	20.910
Ventilated wards of London hospitals	
1. Day	20.920
2. Midnight	20.886
3. Morning	20.884
Carbon dioxide:	ppm
Remote locations	335
London streets, summer	380
London parks	301
On the Thames at London	343

Source: Smith, R.A. (1872) *Air and Rain*, London.

actively engaged in a wide range of research projects which culminated in
the publication of his book *Air and Rain* in 1872, containing the results of
the first air pollution monitoring network in England.[5] Gas analyses were
restricted to oxygen and carbon dioxide, because at the time there were no
established techniques for determining the concentration of trace gases in
the air. Smith overcame this problem by analysing the dust and soot or the
pollutants brought down in the rain-water by collecting it in large gauges.
About the year 1869 Smith arranged for rain-water samples to be collected
from sites all over the British Isles. The collection and analysis of these
went far beyond anything that was required of the Inspectorate by the
Alkali Acts.

Some of Smith's analyses of the atmosphere for oxygen and carbon
dioxide are given in Table 7.1. It can readily be seen that the oxygen
concentration, even in the most confined situations, changes little. Small
wonder that Cavendish, with his extremely primitive apparatus a century
earlier, was not able to find any variation in oxygen with weather. Smith
was also able to show that candles generally went dim when the oxygen
concentration fell to about 18.5 per cent, so the complete oxygen removal
described in 'The Doom of London' could not occur. Smith is careful to
point out that it is not the very small change in oxygen that is important,
'but supposing its place be occupied by hurtful matter, we must not look on
the matter as too small'. Much of the missing oxygen would be accounted
for by carbon dioxide and, as can be seen in the table, the changes in this gas
are proportionately much larger.

Smith's analyses of rainfall composition stretched in geographical extent
from the Hebrides to Southern Ireland. These were usually spot measure-

TABLE 7.2 Composition of rainfall at various locations as determined by R.A. Smith

	Chloride mmol/l*	Sulphate mmol/l	Ammonia mmol/l
Scotland (coastal)	0.36	0.06	0.04
Scotland (inland)	0.1	0.02	0.03
London	0.04	0.2	0.2
Liverpool	0.29	0.4	0.32
Manchester	0.16	0.46	0.38
English towns	0.24	0.35	0.3

Source: Smith, R.A. (1872) *Air and Rain*, London.
Note: *millimoles per litre

ments, but in some urban measurements, notably those made in Manchester, records for a number of years survive. The rainfall was analysed for chloride, sulphate, nitrate, ammonia and a number of other substances. Only the results for sulphate, chloride and ammonia are considered here and summarized in Table 7.2. On the sea coast the ratio of sulphate to chloride in the rainfall was quite low, that is a large amount of chloride was present. In fact it approaches the ratio found in sea-water, suggesting that the maritime rainfall is essentially dilute sea-spray. Inland, particularly in large cities, the amount of sulphate is greatly enhanced. Smith was in no doubt that the source of the sulphuric acid in urban rainfall was the sulphur-containing gases given off in the combustion of coal. These early results mark the beginning of the monitoring that was required to understand the chemical changes that were occurring in the urban atmosphere.

The beginnings of rainfall chemistry

Rainfall composition was of interest to scientists long before Angus Smith became concerned with the changes that arose through anthropogenic activities. Coloured dusts in rain and the presence of sea-salt have evoked comment since classical times. The English scientist William Derham noticed that in the aftermath of the great storm of 1703 the grass had become so salty that the sheep refused to eat it for some time.[6] The atomic theorist John Dalton (1766–1844) showed that the concentration of salt in Manchester rainfall was much higher when the wind blew more directly from the sea.[7] The possible incorporation of sea-spray in the rainfall collected at coastal sites was studied for a short period by the British Rainfall Association. In the 1860s the association made kits available to their members so that they could determine the chloride concentration in rainfall. Symons, the President of the Association, became disillusioned with the venture after a short time, but not before it had been established that, even at coastal stations, sea-spray made only a small contribution to the total rainfall catch.[8] The Rivers Pollution Commission also showed some interest in rainfall for a brief period in the 1860s. They were concerned about the composition of rainfall caught for drinking purposes. Their analyses justified their concern, as even at sites over 25 miles from any large town, a surprising degree of contamination was to be found.[9]

The most significant advances in monitoring came about through agricultural studies. Ever since Roman times the agricultural importance of rainfall composition had been known.[10] When Samuel Johnson said that the 'rain is good for vegetables' he was summarizing a notion that had persisted from the earliest times.[11] Rain was thought to be superior for irrigation of crops because of the compounds it contained. Sir Kenelm Digby correctly thought that these compounds were nitrogenous. However, it wasn't always true that rain aided the growth of crops. The classical writer Pliny[12] described the vegetation damage that was possible when the rain became excessively saline, so it is hardly surprising that some of the earliest agricultural measurements were concerned with the chlorine content of rainfall. In addition to this, chloride in rainfall was relatively easy to determine. The first analysis of a full year's rainfall, albeit rather inaccurate, in the British Isles is for chloride and it was made at a near-coastal location at Penicuik in Scotland in 1843.[13] Some theorized that plants such as the marigold that had a marine ancestry ought to benefit from the addition of salt to the soils in which they grew. This idea was soon discarded and it became evident as time went on that rainborne sea salt had little effect on most crops except right at the coastal margin, and there the effect was probably harmful.[14]

The most active interest in precipitation composition in the last century revolved around the nitrogenous components. Here the great agricultural theorists of the time met head on, and reputations crumbled. Justus von Liebig, in his influential and widely publicized writings, rejected the old ideas of plant nutrition via humus and put forward new concepts.[15] He thought that ammonia was taken up from the air in much the same way as carbon dioxide. The implications of such a theory were enormous. It meant that the addition of fertilizers to the soil was unnecessary. Rainfall was also thought to make a significant contribution to the soil nitrogen and he claimed that some 27 kg were deposited on every hectare in a year, sufficient (it would have seemed) for many crops. The controversy this sparked off initiated much of the rainfall analysis that took place in the seventy years that followed.

In England in the 1850s the early analytical work on this problem was given to the young agricultural chemist J. T. Way. It could not be done at

FIGURE 7.2 Gauge for collecting rainfall for analysis at Rothamsted

the experimental station at Rothamsted as the fields were fully occupied with other studies. The work proceeded smoothly, and it soon became evident that von Liebig's suggested nitrogen levels in rain were much too large.[16] By 1861 it was clear that non-leguminous crops could not assimilate nitrogen from the air and the quantities of nitrogen brought down in the rain were too small to supply the needs of most agricultural crops. After such a definite answer, the subject might well not have received any more attention, but in 1870 the famous large rain gauges of a thousandth of an acre area were erected at Rothamsted, to help with studies of nitrogen loss in percolating drainage waters. Naturally this study required the analysis of the rain-water. The site was also adopted by Frankland of the Rivers Pollution Commission. After installation the large rain gauge was used for chemical determinations through to 1916, until the death of N. H. J. Miller, the chemist who tended the experiments all those years. After such a lengthy study it was tragic that his death occurred just before he had finished the massive report that was to summarize the findings of more than sixty years' work on precipitation chemistry at Rothamsted.[17] Fortunately much of the early data still survives so it can be examined and compared with recent measurements made at the same location.[18] The longest series of measurements in the Rothamsted record

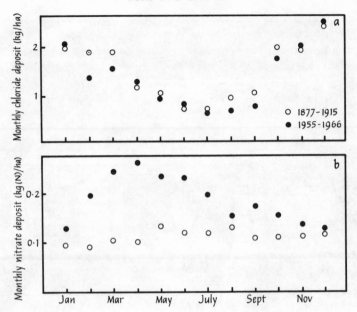

FIGURE 7.3 Changes in the seasonal composition of rainfall at Rothamsted for 1877–1915 and 1955–66

are for the chloride, ammonia and nitrate in rainfall. The sulphate analyses
are sadly more limited, although the few that are available are interesting to
compare with the modern determinations. Between 1881 and 1887 some
7.8 kg of sulphur was deposited on each hectare (ha) at Rothamsted. In the
measurements made between 1955 and 1966 this had increased to
12.2 kg/ha, which may well be due to the increased amounts of sulphur
present in the air over the British Isles. Unfortunately it is not easy to be
certain about the reliability of the early sulphur analyses.

In the case of the chloride, ammonia and nitrate analyses we are much
more fortunate, because here we have monthly analyses in addition to the
annual totals. In Fig. 7.3(a) we can find a nearly perfect overlap between the
modern figures and those obtained from the early data. It is not simply that
the amount of chlorine deposited has remained constant, but the seasonal
distribution has not varied either. No change in the chloride levels in
rainfall are to be expected at a site like Rothamsted, where the main source
of chloride in rainfall should be from sea-salt driven far inland. Even today
man's contribution is much smaller than this natural source of chloride.
Particularly evident in the seasonal distribution is the higher deposit of
chlorine in the winter months which comes about as a result of winter
storms.

If the seasonal distribution of nitrate deposit is examined (see Fig.
7.3(b)), a difference between the modern and old values is quite evident.

Not only is more deposited each year in modern times, but the distribution has changed. Last century there was little seasonal variation, but nowadays there is a peak in nitrate deposit in spring. The overall increase in nitrate may be due to increased pollution emission from London as combustion processes, especially those in the internal combustion engine, produce nitrogen oxides which can then be oxidized to nitric acid which will appear in rain as nitrates. This source may be more important in spring because of increased atmospheric photochemistry or different weather patterns. It is also possible that the change in nitrate may indicate a change in agricultural practice and increased activity within the nitrogen cycle as a result of the application of vastly increased amounts of fertilizers over the last century.

Monitoring within the metropolis

The earliest measurements of the composition of rainfall in London come from measurements made by R. A. Smith which were included in his book *Air and Rain*. The stations were maintained for only very short periods and the results suffered from the fact that samples were often collected from different seasons over varying lengths of time. This makes any comparison with later, more systematic surveys nearly impossible. The data collected by Smith came from a large number of sites (mostly fire-stations), so if we are willing to accept the limitations to its accuracy, then it is possible to construct a contour map of the sulphate concentration in the rain samples around London. The determinations made by Smith were largely on samples that were collected in the first half of 1870 and so give us a crude idea of the distribution of the sulphate deposited in London rain more than a hundred years ago. As expected, the highest concentrations are to be found in central, densely populated areas of London. There seems to be some elongation of the area of high concentration in an east-west direction which is no doubt aided by the prevailing wind (Fig. 7.4). The collection technique and long exposure times used by Smith mean the concentrations are not comparable with modern measurements. However Smith was able to see that most of the sulphur was not deposited within the boundaries of cities, but was transferred into rural areas.

Dr W. J. Russell was the next chemist to take an active interest in London rain. While his sampling network was restricted to just a few sites, the samples were collected at fairly frequent intervals between 1882 and 1884.[19] His record does not show how much rain was sampled each time but it is possible to establish this from London rainfall records of the period. Combining these two sets of data enables us to make an estimate of the annual deposit of sulphate and chloride in London for an early date (see Table 7.3(a)), decades before official monitoring programmes began. These early measurements show sulphates rather higher than those we

TABLE 7.3 Annual wet deposition of chloride and sulphate anions (expressed
as grams of the element per square metre) (a) as estimated from the rainfall
analyses made by W.J. Russell from October 1882 to March 1884 and (b) by
the DSIR for 1956–7

(a) 1882–4

	St Bartholomew's Hospital EC	Hamilton Tce NW	Shackwell Green
Chloride	10	7.5	3.5
Sulphate	7.3	4.2	3.8

(b) 1956–7

	Central London 4 sites	Hampstead 2 sites	Stoke Newington 1 site
Chloride	4.6	6.5	4.0
Sulphate	5.2	3.1	3.3

Sources: Data from Russell, W.J. (1884) 'On London rain', Appendix I, Monthly Weather
Report, (April).

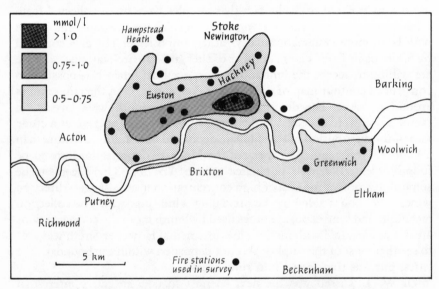

FIGURE 7.4 The amount of sulphate deposited in rain around London,
1869–70

would expect today (see Table 7.3(b)). There were no further analyses of
London rainfall until 1910, when Dr H. A. Des Voeux of the Coal Smoke
Abatement Society inspired a small trial network.[20] These analyses
represented a broadening of the interests of the Coal Smoke Abatement
Society (1889). Up until that time it had been concerned mainly with the

FIGURE 7.5 Dr W. J. Russell of St Bartholomew's Hospital analysed the composition of London rain and air in the last century

observation of black smoke emissions and miscellaneous activities such as running courses for stokers.

The work of Russell and the Coal Smoke Abatement Society was the first step in the development of the deposit gauge[21] (see Fig. 7.6). The gauge is essentially a large rain gauge with a screen to prevent birds from contaminating the collection. Every month the bottle below the funnel is replaced with a new one and the contents of the previous month's sample analysed. Analyses typically cover chloride, sulphate, ammonia, calcium, total undissolved matter, tarry matter, ash, acidity or alkalinity and, later, pH. The initial analyses instigated by Des Voeux were carried out by the laboratory of the medical journal *The Lancet*. They were restricted to the analyses for chloride, sulphate and ammonia, some calcium analyses and the measurements of the solid material. This study was the forerunner of an

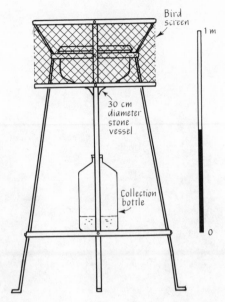

FIGURE 7.6 The gauge developed, early this century, to determine the deposit of airborne soot

extensive network of gauges that was to come under the overall supervision of the Advisory Committee on Air Pollution.

Advisory Committee on Air Pollution

In March 1912 the International Smoke Abatement Exhibition was held in London under the auspices of the Coal Smoke Abatement Society. There were not merely exhibits, but a number of technical papers were presented. Some of these emphasized the need for a systematic nationwide investigation into atmospheric pollution. Soon after the exhibition, the Committee for the Investigation of Atmospheric Pollution came into being under the chairmanship of Sir Napier Shaw, then Director of the Meteorological Office.[22] The committee was set up on a purely voluntary basis, but *The Lancet* offered both its publishing and its laboratory facilities. The early membership list is most impressive. It seems to include a large number of people whom we would regard as important figures in early twentieth-century air pollution science. The members and some of their contributions are listed in Table 7.4. The group was to engage in the scientific investigation of air pollution, rather than to press for political or legislative changes. In this way they differed from the numerous bodies that had previously tackled the problems of London's polluted atmosphere.

TABLE 7.4 The early members of the Committee for the Investigation of Atmospheric Pollution

---

Chairman: Sir Napier Shaw, who is remembered as the first Director of the Meteorological Office and the author of a large number of books and papers on meteorological topics.

Secretary: Dr J.S. Owens of the Coal Smoke Abatement Society, who did much research on air pollution in London over the decades that followed. He wrote an important book, *The Smoke Problem of Great Cities*, with Shaw.

J.F. Cave, a past president of the Royal Meteorological Society and author of *The Structure of the Atmosphere*.

J.G. Clark of the Meteorological Office, who was appointed to the sub-committee on deposit gauges.

Prof. J.B. Cohen of the University of Leeds and author of a number of papers on air pollution and its effects on vegetation in and around the city.

Dr H.A. Des Voeux, Hon. Treasurer of the Coal Smoke Abatement Society.

Dr Hawkesley, Assistant Medical Officer of Health, Liverpool.

J.B.C. Kershaw, who was a member of the Hamburg Smoke Abatement Society and known as the co-author of *Smoke Prevention and Fuel Economy*.

Dr E.J. Russell, Director of Rothamsted Experimental Station, who had been responsible for analysing the data collected by Miller after his untimely death.

E.D. Simon of the Smoke Abatement League of Great Britain.

Bailie W. Smith of the Sub-Committee of the Glasgow Corporation, who was the Convenor of the Air Purification Sub-Committee on deposit gauges.

S.A. Vasey of *The Lancet*.

F.J.W. Whipple, Superintendent of the Instruments Division, Meteorological Office.

A sub-committee consisting of J. G. Clark and Bailie Smith was responsible for the design of the standard deposit gauge. The gauge used in *The Lancet* study of 1910–11 had been made of sheet iron painted with enamel. This material proved to be satisfactory only for relatively short exposures in the corrosive atmospheres of the city. Initially a glass wool plug was placed in the neck of the funnel to filter out the solid impurity in the rain-water, but the glass wool also dissolved in the rain-water! The standard gauge had a circular bowl to replace the initial square one and was of cast iron with a coating of vitreous enamel, rather than enamelled sheet metal. Although the standard gauge was meant to have an area of exactly 4 square feet, the enamelling process imposed such strains on the vessels that they ceased to be truly circular. The problems caused by this distortion were overcome by the calibration of each gauge with a special gauge factor to allow for deviations from a circle. This practice of using a gauge factor still persists in the collection of atmospheric deposit. A wire screen was placed around the gauge to prevent birds from perching on the edge. No precautions taken against birds ever seem to be completely adequate and

FIGURE 7.7 Sir Napier Shaw, chairman of the Committee for the Investigation of Atmospheric Pollution and Director of the Meteorological Office

methods of preventing them from perching on apparatus in the field still continue to worry scientists who collect precipitation samples. The gauges were to be distributed to nineteen cities and towns over the British Isles and by April of 1914 twelve gauges were already in operation. In time the gauge underwent further evolution, the bowl being reduced in size and subsequently made of enamelled stoneware.

By the second year of its existence the committee had changed its name to the Advisory Committee on Atmospheric Pollution. The results obtained from the network of gauges it operated were printed in reports published each year by the Meteorological Office. The committee had been awarded £500 a year by the Department of Scientific and Industrial Research on the condition that the Meteorological Committee, which at that time controlled the work of the Meteorological Office, accepted responsibility for the expenditure of the grant. Although the Meteorological Office published the annual reports, monthly air pollution data still appeared in full in *The Lancet*. This arrangement continued for many years until the air pollution monitoring fell directly under the control of the Department of Scientific and Industrial Research (DSIR).

As the years went by the number of deposit gauges at the disposal of the DSIR increased enormously. In the Glasgow area the analysts were often troubled by the presence of beer-sugars turning up in city gauges. At other sites urine was also found. During the Second World War occasional entries describing massive deposits of solid material are annotated with comments that the source was largely dust raised by enemy action.[23] However, with the increasing sophistication of instruments in the 1950s there were many who felt that the deposit gauge network needed to be replaced by more modern devices. One of the critical problems with the deposit gauge was that even when carefully sited its accuracy was no better than ± 20 per cent.[24] It was also true that the deposit often reflected materials that were largely of very local origin and did not give a representative picture of the pollution over a wider area. In the early 1970s Warren Spring Laboratories, the modern heirs to the network, anxious to move to surer techniques, showed declining interest in the deposit gauge and the long records that had accumulated.[25]

Despite the pressures to abandon deposit gauge measurements, they have continued in a number of cities through the 1970s. In London a few enthusiasts ensured that the Greater London Council continued to monitor the deposit within the city. This means that a record of more than sixty years now exists for some sites in London. Through judicious use of some of the earlier data it is possible to assemble the seventy years' record that is shown in Fig. 7.8. No fear about the absolute accuracy of the record can fail to show the great changes over the last century that have led to a decline in the amount of soot deposited in the London area. The decrease in sulphate deposits in the century-long record, while not so clear, is evidence of great improvement also.[26]

Measurement of suspended particulates

The deposit gauge provided a simple way of sampling the urban atmosphere, but really provided only a sketchy idea of its composition, because just what fraction of the components of the atmosphere were washed out in showers and what fraction remained in the air was unknown. The deposit gauge was just a way of measuring 'sootfall' and did not really measure air pollution at all. Early workers realized this and in the 1880s Russell made some measurements of the actual particulate loading of the

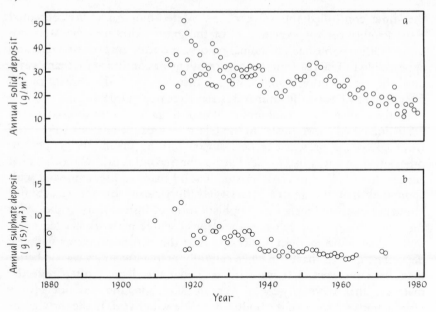

FIGURE 7.8 The record of total undissolved solid deposit and sulphate
in London

London atmosphere by sucking large volumes of air through filters.[27] The
technique was extremely tedious and only a few measurements were made,
but they are certainly interesting and indicate that the particulate loading of
the London air was high in the last century. Russell's data is compared with
more recent measurements for a range of conditions in Table 7.5.
It is particularly interesting to see how estimates of the smoke vary with
weather. The mean modern values are much lower but in periods of intense
fog they could exceed 4000 $\mu g/m^3$. Note the lower values found at the
weekend.

In the absence of air-sampling equipment people sometimes used quite
ingenious sources for information on the composition of the material
suspended in the atmosphere of London. In one case after the turn of the
century the solid deposits that accumulated on ventilator shafts were
analysed.[28] In another sampling exercise the deposit on glass panels of a
roof in Chelsea was analysed. However, the sample in this latter case is a
little akin to that collected by the deposit gauge and does not really tell us
much about the concentration of material in the atmosphere.

This problem of measuring suspended material in the atmosphere began
to receive the attention of the Advisory Committee on Atmospheric
Pollution as soon as it had the network of deposit gauges running
smoothly. The aim was to develop a technique that would enable the
concentration of particulate matter in the London atmosphere to be
determined rapidly without the aid of a skilled chemist. The method

TABLE 7.5 Measurements of the concentration of soot in the atmosphere of London

	Fine weather	Dull weather	Foggy weather
1885			
Smoke ($\mu g/m^3$)	120	360	860
Severe fog of December, 1962			
Smoke ($\mu g/m^3$)			> 4000

	Summer		Winter	
	weekdays	*weekends*	*weekdays*	*weekends*
1956–7				
Smoke ($\mu g/m^3$)	84	64	185	173

Sources: Russell, W.J. (1885) 'On the impurities of London air', *Monthly Weather Report*, (August); *The Investigation of Atmospheric Pollution*, 30th Report, HMSO, London (1959); *The Investigation of Atmospheric Pollution 1958–1966*, 32nd Report, HMSO, London (1967).

adopted consisted of drawing air through a small circle of filter paper and noting the amount of darkening that resulted from the capture of particulate material. A great deal of time and trouble had to be taken to calibrate the degree of darkening against the amount of suspended particulate on the filter papers. Once this was done, however, it provided a very rapid method of analysis. By 1918 a device was available which ran automatically, making a determination every 15 minutes. An instrument of this type was installed at Kew in the 1920s.[29] The measurement of particulate concentration in the atmosphere by the 'smoke shade' on white filter paper is still a very popular method because of its simplicity. However, the technique is susceptible to error over long periods as the actual nature and hence colour of the smoke changes.[30] In London, for instance, this change in colour has come about as the major source of smoke has shifted from coal to petroleum products. The degree of error can vary between threefold overestimates and sixfold underestimates of the actual concentration of smoke particles in the atmosphere. Naturally this can be overcome by constant recalibration, but it has become common practice to use high-volume samplers which suck such large amounts of air through large glass-fibre filters so that they can be weighed directly.

Sulphur dioxide

The early studies of the Advisory Committee on Atmospheric Pollution concentrated largely on the suspended particulate materials (smoke) and the deposited material. The deposit gauges gave some hint as to the actual

TABLE 7.6 Early estimates of the sulphur dioxide concentrations in English towns and cities.

Location	Sulphur dioxide $\mu g/m^3$ (as SO_2)
Manchester	2930
London	2180
Buxton	1950
Didsbury	1746
Blackpool	620

Source: Mayberry, C.F. (1895) *J. Am. Chem. Soc.*, 17, 105.

chemistry of the air pollution in London, but progress required more detailed analysis of the gaseous components of the air. Damage to building stone and vegetation was already known to arise through the action of sulphuric acid, the product of the atmospheric oxidation of sulphur dioxide, so there were good reasons for examining the sulphur dioxide concentration in London's atmosphere. There were a few determinations of the concentration of sulphur dioxide in the air of various cities of Europe early in the nineteenth century, but concentration figures are all so high that it is hard to believe they are reliable. Table 7.6 shows values obtained in a number of English towns and cities late last century, and although they also seem too high to be representative of anything but the most polluted days of the year, they do enable us to compare sulphur dioxide pollution in various places.[31]

Some progress was made in developing a simple method of measuring atmospheric sulphur dioxide in 1932. The Building Research Station was able to estimate the amount of gas absorbed by an exposed solid surface. They chose lead peroxide (PbO_2) as the surface because it would react with sulphur dioxide according to the reaction

$$PbO_2 + SO_2 \rightarrow PbSO_4$$

After a given period of exposure the surface could be analysed and the sulphur dioxide absorbed expressed in terms of its deposition on a certain area of exposed lead peroxide. The procedure was found to be both simple and inexpensive and gave a reasonable measure of the amount of sulphur dioxide deposited. This information was obviously of great interest to the scientists at the Building Research Station, who were concerned with the damage that deposited sulphur dioxide did to paint, stone and iron work. The device was so successful that it became widely used in monitoring in general. However, it has not proved as successful when used in estimating actual concentrations of sulphur dioxide in the air. It is difficult to convert the measurements from the amount of sulphur deposited into an estimate of the mean concentration of sulphur dioxide in the air over the period of exposure.[32]

A more satisfactory technique is to draw a known volume of air through a solution which can then be analysed for dissolved sulphur. In general about 2 cubic metres of air is drawn through about 30 cubic centimetres of a dilute hydrogen peroxide solution at pH 4.5. The solution of hydrogen peroxide oxidizes any sulphur dioxide in the gas stream to sulphuric acid:

$$H_2O_2 + SO_2 \rightarrow H_2SO_4$$

The amount of sulphuric acid can then be determined by titration with a solution of alkali to bring the solution back to a pH of 4.5. The analytical procedure suffers from the problem that other acidic or alkaline gases in the atmosphere, such as ammonia, hydrochloric acid and the nitrogen oxides may interfere by changing the acidity or alkalinity of the solution. In more recent times it has become usual to analyse for the amount of sulphur in the solution directly by a colorimetric method or electrochemical methods.[33]

Floral and faunal studies

Londoners had long been aware of the influence of the city air on their gardens, but scientific studies didn't start until the late nineteenth century.[34] In addition to the scientific work, horticultural societies in London maintained an active interest in the effect of smoke on plants. The increasing spatial extent of polluted air around London meant that some seedsmen and florists who had traditionally had gardens quite near the centre of the city were obliged to shift into the country at the turn of the century. A most extensive study of the effect of town air on vegetation commenced on Manor Farm at Garforth near Leeds in 1906. The study was part of long-term research in air pollution and agriculture which was maintained at the University of Leeds.[35] The work examined the role of sulphur dioxide and of the acids produced in its oxidation in damaging vegetation. The researchers were also interested in the importance of soot in cutting off sunlight and the effect of soot deposits on plant leaves. Soot deposits can block the stomata and inhibit photosynthesis. The report from the project concluded with, among others, the following statements:

(i) the atmosphere in and around large industrial cities has a marked deleterious effect upon plant growth;
(ii) while the effects are dominant in the industrial quarters of the city, winds disperse the pollutants over large areas;
(iii) the rain falling through the polluted air brings down a large amount of pollution.

It became evident to the group working on this project (Charles Crowther, Arthur Ruston and J. B. Cohen) that plants were an excellent

TABLE 7.7 Sulphur dioxide concentrations in Epping Forest as estimated from lichen surveys

Date	Investigators	Sulphur dioxide winter mean $\mu g/m^3$
1784–96	Edward Forster	30
1865–68	Crombie	40
1881–82	Crombie	50– 60
1909–19	Paulson and Thompson	60– 70
1969–70	Hawksworth	70–120

Source: Hawksworth, D.L., Rose, F. and Coppins, B.J. (1973) 'Change in the lichen flora of England and Wales attributable to pollution of the air by sulphur dioxide', in Ferry, B.W., Baddeley, M.S. and Hawksworth, D.L., *Air Pollution and Lichens*, Athlone Press, London, 330–67.

indicator of air pollution. In a sense they provided, free of charge, a kind of biological monitoring network. The abundance of species and their health in a given area can be used to gauge air quality. Although this was suggested as early as 1915, the most impressive applications of this technique have not been made until quite recently. Some lichen species are particularly sensitive to sulphur dioxide, so it has been possible to use their distribution throughout the British Isles to draw up maps of the distribution of sulphur dioxide.[36] More relevant to the historical question is the study of the variation in the lichen population of Epping Forest since the eighteenth century, when the earliest lichen survey was made.[37] Epping Forest is to the north-east of London and the city boundaries have come ever closer to the forest. In such a situation a gradual increase in sulphur dioxide levels is to be expected. The data presented in Table 7.7 shows the gradual increase in mean winter sulphur dioxide concentrations estimated from the lichen populations. There is some evidence that the lichen may be moving back into the City of London, but in general it is the acid-tolerant varieties that prove most effective at recolonization.

Perhaps a more remarkable phenomenon had been the evolutionary changes imposed on some species by the increasing pollution of the British Isles. The best-known example of such an evolutionary change is the increase in melanic forms of various butterflies. Obviously in the darkened regions of the Midlands, black was the most suitable colour for camouflage; so sure enough, black types of moths began to be found more frequently in the last century,[38] and there is evidence of increasing melanism in butterflies in the London area since that time.[39] The distribution of melanic forms of *Biston betularia* bears a remarkable similarity to the distribution of fuel use, and hence air pollution, within the British Isles.[40]

More recent evidence of adaptation to polluted environments has come from the studies of perennial rye grass populations in the Merseyside area.[41] Thirty-six individually identified clones of each population were exposed to either a background control sulphur dioxide concentration of 35 $\mu g/m^3$ or to a high concentration of 650 $\mu g/m^3$. The populations had been taken from both urban and rural sites and when the growth in the high sulphur dioxide concentration experiments was compared to the background set, it was found that the rural population appeared to be less successful at growing in air polluted with sulphur dioxide. This suggests that long exposure to the polluted air of the North has caused even the grasses to evolve into more pollution-tolerant varieties.

Throughout this chapter we have seen that urban air pollutants and their effects were ever more carefully monitored. This was paralleled by a growing understanding of the nature of the pollutants in urban air. However, legislative changes and abatement practices did not keep pace with the implications of this increased knowledge and, almost inevitably, tragedy followed.

Notes

1. As has been emphasized, scientists had long been aware of the sulphurous content of coal smoke. The mistake that tended to be made in the early nineteenth century was the view that it was too small to be important. Buchanan, W. M. (1857) *Smoke Nuisance Question*, Griffin & Co., London, points out that very little sulphur is present in coal. This may seem a crude notion, but it really did survive in the minds of many in the field until the 1950s. After all, Marsh's classic book on the problem of coal and the atmosphere is simply titled *Smoke* (Marsh, A. (1947) *Smoke*, Faber & Faber, London).

2. *Report of the Committee (Lord Derby's) on the injury resulting from Noxious Vapours evolved in certain manufacturing processes, and the Law relating thereto; Evidence, Appendix and Index* (1862). An article on the operation of this legislation is given in MacLeod, R. M. (1965) 'The Alkali Acts administration, 1863–84: the emergence of the civil scientist', *Victorian Studies* 86–112.

3. Kingzett, C. T. (1877) *The History, Products, and Processes of the Alkali Trade*, Longman, Green, London; Lungo, G. (1891–6) *A Theoretical and Practical Treatise on the Manufacture of Sulphuric Acid and Alkali*, Gurney & Jackson, London.

4. Ashby, E. (1975) 'The politics of noxious vapours', *Glass Technology*, 16, 60–7.

5. Smith, R. A. (1872) *Air and Rain*, London.

6. Derham, J., 'Observations of the late great storm', *Phil. Trans. Roy. Soc. Lond.* ab. v 60 (1704).

7. Dalton, J. (1824) 'Saline impregnation of rain which fell during the late storm, December 5th 1822', *Memoirs Lit. and Phil. Soc. Manchester*, 4, 324–31 and 363–72.

8. Symons (1886) 'Detection of sea spray when mingled with rain', *British Rainfall*, 11–12.

9. Frankland, E. (1874) *Sixth Report, Rivers Pollution Commission*.

10. Vitruvius, *De Architectura*, VIII, 2.

11. Johnson, S. in *Boswell's Life of Johnson*, 14 July 1763.

12. Pliny, Lib. xxxi, c. 29, 'Ter accidit in Bosporo, ut salsi deciderent necarentque frumenta.'

13. Madden, H. R. (1843) 'On the advantages of extended chemical analysis to agriculture', *Trans. Highland Agric. Soc.*, 14, 568–86.

14. Matson, M. (1876) 'Salt in rainwater', *Agric. Student's Gazette*, 1, 132; Church, A. H. (1877) 'Salt in rainwater', *Agric. Students' Gazette* 2, 14.

15. von Liebig, J. (1843) *Chemistry and its Applications to Agriculture and Physiology*, Taylor & Walton, London.

16. Way, J. T. (1855) 'The atmosphere as a source of nitrogen to plants', *J. Roy. Agric. S. England*, 16, 249–67.

17. However, the work of Miller survives in two fairly complete papers: Miller, N. H. J. (1905) 'The amounts of nitrogen as ammonia and as nitric acid and of chlorine in the rain water collected at Rothamsted', *J. Agric. Sci.* 1, 280–303, and Russell, E. J. and Richards, E. H. (1919) 'The amount and composition of rain falling at Rothamsted', *J. Agric. Sci.*, 9, 309–37.

18. As part of the European Air Chemistry network the results appeared regularly in *Tellus* from 1955–1966. Since then they have continued with somewhat less regularity; for details see Brimblecombe, P. and Pitman, J. I. (1980) 'Long term deposit at Rothamsted, England', *Tellus*, 32, 261–7.

19. Russell, W. J. (1884) 'On London rain', Appendix I, *Monthly Weather Report*, (April).

20. Anon. (1912) 'The sootfall of London: its amount, quality and effects', *The Lancet*, 47–50.

21. 'Atmospheric pollution', *Nature*, 94 (1914), 433–4.

22. Some idea of the early history of the committee can be gained from its earliest annual reports.

23. Occasional humorous incidents can unfortunately only be found by reading many thousand, now yellowing, pages of analyses.

24. Craxford, S. R., speaking on the B.S. 1747–1951 deposit gauge at the Annual Conference of the National Society for Clean Air, 1960.

25. *Deposit Gauge and Lead Oxide Observations*, Department of Trade and Industry, Warren Spring Laboratories.

26. I am indebted to Bill Culley of the former Greater London Council (Scientific Branch) for his assistance in gathering this data. Brimblecombe, P. (1982) 'Long term trends in London fog', *Science of the Total Environment*, 22, 19–29; Brimblecombe, P. (1982) 'Trends in the deposition of sulphate and total solids in London', *Science of the Total Environment*, 22, 97–103.

27. Russell, W. J. (1885) 'On the impurities of London air', *Monthly Weather Report*, August.

28. *Committee for the Investigations of Atmospheric Pollution*, 1st report, April 1914–March 1915; Lewes, V. B. (1910). 'Smoke and its prevention', *Nature*, 85, 290–4. The analyses are as follows:

Component	Deposit on glass roof in Chelsea (1910) (%)	Ventilator filters in central London (1914) (%)
Carbon	39.0	35.5
Hydrocarbons	12.3	
Organic bases	1.2	
Tar		1.5
Calcium		2.8
Ferric oxide		2.5
Magnetic fraction	2.6	
Alumina		8.3
Sulphate	4.3	4.5
Ammonia	1.4	0.9
Silica	31.2	38.0
Fibres		1.0

Despite differences in these analyses and in the classification of the components, carbon and siliceous materials dominate.

29. Thornes, J., 'London's changing meteorology', in Clout, H. (ed.) (1978) *Changing London*, University Tutorial Press, London.

30. Ball, D. J. and Hume, R. (1977) 'The relative importance of vehicular and domestic emissions of dark smoke in greater London in the mid 1970s, the significance of smoke shade measurements, and an explanation of the relationship of smoke shade to gravimetric measurements of particulate', *Atmos. Env.* 11, 1065–73.

31. Ladureau, A. (1883) *Ann. Chim. Phys.*, 29 (5), 427, for a rather high measurement of sulphur dioxide in cities.

32. *The Investigation of Atmospheric Pollution*, (1960) 31st Report, HMSO, London.

33. Katz, M. (1969) *Measurement of Air Pollutants*, WHO, Geneva includes many analytical methods, but that for sulphate is colorimetric. Currently the measurements in London are made by the British Standard procedure and an electrochemical method; Scwhwar, M. J. R. and Ball, D. J. (1983) *Thirty Years On*, GLC, London.

34. Voelcker, A. (1864) 'On the injurious effects of smoke on certain building stones and on vegetation', *J. Soc. Arts.* 12, 146–51; Slater, A. (1875) 'Note on a wood damaged by gases from calcining iron stones'. *Trans. Scot. Abor. Soc.*, 8, 184–5.

35. Crowther, C. and Ruston, A. G. (1911) 'The nature, distribution and effects upon vegetation of atmospheric impurities in and near an industrial town', *J. Agric. Sci.*, 4, 25–55.

36. Hawksworth, D. L. and Rose, F. (1970) 'Qualitative scale for estimating sulphur dioxide pollution in England and Wales using epiphytic lichen', *Nature*, 227, 145–8.

37. Hawksworth, D. L., Rose, F. and Coppins, B. J. (1973) 'Change in the lichen flora of England and Wales attributable to pollution of the air by sulphur dioxide', in Ferry, B. W., Baddeley, M. S. and Hawksworth, D. L. *Air Pollution and Lichens*, Athlone Press, London, 330–367.

38. Kettlewell, H. B. D. (1973) *The Evolution of Melanism*, Clarendon Press, Oxford.
39. Mera, A. W. (1926) 'Increase in melanism in the last half-century', *London Naturalist*, 3–9.
40. Kettlewell, H. B. D., 'A survey of the frequencies of *Biston betularia* (L.) (Lep.) and its melanic forms in Great Britain', *Heredity*, 12 (1958), 51–72.
41. Horsman, D. C., Roberts, T. M. and Bradshaw, A. D. (1978) 'Evolution of sulphur dioxide tolerance in ryegrass', *Nature* 276, 493–4; Bradshaw, A. D. and McNeilly, T. (1981) *Evolution and Pollution*, Edward Arnold, London.

8

The Great Smog and after

The influential and productive work of the smoke abatement societies and the activities of a few enlightened parliamentary committees in the last century promised rapid improvement in the control of air pollution. It is evident that the air over London did become less polluted in some particular localities, but the advances were considerably slower than any of the Victorian activists would have hoped. It was more than just administrative apathy: two great wars and an economic depression helped to keep clean air from becoming a reality.

The air pollution in London continued to be severe in the early twentieth century despite the slight decline in the frequency of London fog. Long exposure to polluted atmospheres brought new problems to light. In the railway station at Charing Cross a girder collapsed. Analysis showed that it contained nearly 9 per cent ferrous sulphate which had been produced from continued exposure to sulphurous coal smoke.[1] Investigations in other stations showed the problem was not unique to Charing Cross. The stonework of London's buildings decayed more rapidly than ever at the turn of the century. The degradation of the fabric of many older buildings had been greater in the previous fifty years than it had been in their whole lifetime before that. Such damage affected London's great architectural heritage. Fashions also continued to be affected. Dull-coloured outdoor paints remained the most popular. Indoors, dark-coloured wallpapers were in vogue and the cleaning of curtains was a continuing problem. The decline in popularity of silver plate through the early part of this century is blamed on the tarnishing effect of the urban atmosphere,[2] but then it may merely have been that good servants to clean it were becoming so hard to find.

The increasing traffic of trains and boats created its own pollution problem, although the smoke emissions from both these classes of vehicles had been controlled, to some extent, by various Acts for a considerable time. The grubby state of the London underground tunnels, despite many improvements in recent times, is often blamed on the use of steam trains within the tunnels in the early days.[3] As air traffic developed it was constantly troubled by the high frequency of fogs and the long drift of smoke haze from London. One flying manual reassures the pilot that smoke haze shouldn't be found farther than 70 miles from the city![4] Despite the fact that air pollution bothered aviation so much, aircraft contribute relatively little to the atmospheric pollution of the city even today. Their contribution to the noise pollution over the city is more noticeable.

Even in the present century the solution proposed for both domestic and industrial smoke emissions was a transition to either gas or electricity, or failing this a more efficient method of burning coal. This was meant to apply to both domestic and industrial sources. In industry this was to involve the use of mechanical stokers, and in the home closed fires continued to be strongly advocated. It was realized that the fuel itself could be improved. Coal could be washed to lower both its grit and its pyrites content (the pyrites gives rise to the sulphur dioxide). There was an increasing demand for this and for smokeless fuels (anthracite, coke and semi-cokes such as Coalite) prior to the outbreak of the Second World War. Even after the war one of the great advocates of cleaner air, Arnold Marsh, could say in his book *Smoke*[5] that 'not until standards of air quality are far higher . . . will there be any criticism of the sulphur content of solid smokeless fuel'. As recently as the 1950s the approach to air pollution control, even among some of its strongest proponents, remained essentially one of smoke abatement.

The process of electrification was slow. The rail system, once thought to emit around 2.5 per cent of the country's smoke, has never been fully electrified. In fact for many districts it was the diesel engine rather than the electric motor that replaced the steam engine. In the domestic sector the move to electrification is almost complete, except for heating where the transition has been slower and gas and oil proved efficient competitors. The hearth was retained in the living room long after houses had become 'all-electric'. In the years either side of the Second World War the domestic consumption probably amounted to 25 million tons of bituminous coal each year.

George Orwell was an eloquent defender of *fire worship*.[6] He maintained that although coal was a lot of trouble, dirty and polluting, all these things were 'comparatively unimportant if one thinks in terms of *living* and not merely of saving trouble'. To find Orwell, whose concern for the poor was so keen, defending pollution seems strange, but concern for *living* is not to be neglected. In some senses it is possible to sympathize with him. Pollution can too easily be used to draw attention from other more urgent needs for reform. This happens because often pollution has a simple technological solution, so much easier to envisage than the sociological solutions required to abolish poverty, unemployment or discrimination. An important consideration in the case of Orwell's attitude was the fact that smoke abatement still seemed very much the concern of the industrialists and the intelligentsia. They were trying to impose solutions on people in whose eyes the problem of air quality rated well below that of housing and sanitation. Even today the urban poor may still be far more concerned with their immediate economic problems than those of the air pollution which drifts far outside the city ghettos, to annoy the rich and environmentally conscious.

Despite all the resistance, change was taking place. In the postwar years, the gradual movement back to peacetime prosperity brought with it the desire for more labour-saving devices in the home. The technological changes needed for a less polluted atmosphere were as inexorable as they were slow. Parallel to these came equally gradual improvements in environmental law.

Legislation

At the turn of the century smoke control in London was embodied in the Public Health (London) Act of 1891, which had arisen from the Public Health Act of 1875. The Act did not apply to private chimneys, but other fires and furnaces were required, as far as practicable, to consume all smoke. This sounds very reasonable, but the difficulty was that the Act stated that 'any chimney sending forth black smoke might be deemed a nuisance'. How does one define 'black smoke'? The Chelsea Borough Council sued the Underground Electric Railways of London Ltd to oblige it to reduce the smoke from its power station. This action met with no success, because the generating station was able to claim that their smoke was actually dark brown! In 1909 the London County Council's Public Control Committee strove to have these qualifying words, which crippled the 1891 Act, deleted. However, strong commercial forces rallied against them and the London County Council (General Powers) Act that emerged in 1910 was of only limited use in controlling smoke emission.

Measurements of the quantity of sooty deposit falling onto London began to reinforce the need for further legislation. Jack London, the author of Call of the Wild, gave for the soot deposit a slightly exaggerated figure of 24 tons per square mile per week when he wrote his book on the London poor, People of the Abyss.[7] The use of the unit 'tons' no doubt had a great psychological advantage over the current vogue for gram or milligram quantities of deposit. More carefully collected deposition data began to appear in 1912 after the trial experiments run by the laboratory of The Lancet.[8] It was suggested that 76,000 tons of soot fell on the London administrative county each year. This method of expressing the quantity of sootfall, as the weight over the whole county, presumably had even more psychological impact than that used by Jack London.

In 1913 Gordon Harvey, MP for Rochdale, attempted to get a bill

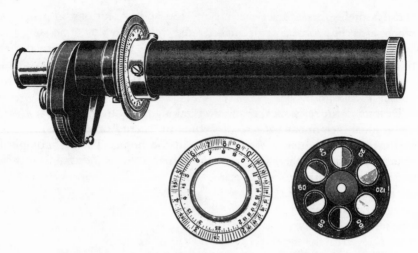

FIGURE 8.1 Messrs B. J. Hall & Co.'s carboscope for determining the
shade of smoke. Smoke was viewed through the telescope and its shade
compared with those of a standard which covered half the field of view

through the Commons which would remove the troublesome word 'black'
from the statute book. He failed. Lord Newton attempted to introduce the
bill to the House of Lords, but it was withdrawn on the announcement that
a committee would be set up to examine the problem. Then it became
caught up in the tangle of the war. However, only four months after the
armistice, the committee was hard at work attempting to take advantage of
the postwar rebuilding programme, but it became evident that the nation
was still attached to its open fires. Nevertheless the report of the committee
did emphasize the need for *all* industry to use the best practicable means to
prevent smoke (*of all colours*). In the coal strikes of 1921, Londoners were
reminded of the joys of clean air, in much the same way as the blockades of
Newcastle in Restoration London had reminded people in John Evelyn's
time.[9]

Throughout the early years of the 1920s and the changes in government
they brought, there were a number of attempts to create workable
legislation that would combat the smoke problem. Laws existed, but
everyone acknowledged that they were crippled by too many exemptions
and weaknesses. At the centre of the legal problem was the difficulty of
defining smoke itself. The early Alkali Acts had not had a problem; they
simply set limits to the amount of hydrochloric acid that could be emitted;
but smoke was not such a simple substance. At one stage a unit called the
'murk' was defined, but has been sadly forgotten! An elegant instrument
was designed for estimating the blackness of smoke (Fig. 8.1). It too has
dropped from the catalogues of scientific suppliers.

The year 1926 saw the passage of the Public Health (Smoke Abatement) bill, a fairly feeble piece of legislation that had suffered much in its passage through the House of Commons. However, the Act that ensued managed to have had the word 'black' removed and placed the onus of proving that the *best practicable means* had been used on the polluter. But the domestic pollution issue was the one over which the House would not budge. Although the smoke abatement lobby now had the electricity and the gas companies, with their vested interests, fighting with them, the Englishman's right to a roaring fire remained sacrosanct. Mr Story Deans, MP said that 'although I may be told that the smoke from my coal fire assists in poisoning the people outside, I prefer that very much to being poisoned myself by a gas fire within my own house'.[10] Despite the inherent selfishness in this kind of attitude it represents the last voice of freedom of an age in which what you burnt in your grate to keep warm was your own business. Eventually the complexities of a modern civilization were to infringe even on that right. Still there was some way to go. 'Best practicable means' was proving no less difficult to define than black smoke. Although some provision for establishing local standards was set up by the Newton Committee it was always possible that some expert might be found to convince the courts that the best practicable means had been used even though emissions of smoke were undesirably high. By 1932, 155 authorities had framed by-laws concerning the abatement of smoke but for the main they related only to black smoke and the duration of its emission.

Seven years later the world was once again plunged into war. The National Smoke Abatement Society leadership fell into the hands of Charles Gandy, who like John Evelyn three centuries earlier perceived that the rebuilding of London after a crisis would offer great possibilities for the creation of smokeless zones. In 1946 the City of London (Various Powers) Act and the Manchester Corporation Act allowed these cities to create smokeless zones. When a draft bill for smokeless zones in Coventry was opposed it was brought in by a crushing majority in a civic referendum. Public opinion was, at least in some areas, now sensitive to environmental issues. The advances in environmental legislation, as in smoke control technology, were more rapid in the post-war years, but all these changes paled before those catalysed by 'The Great Smog' of 1952.

'Killer Smog'[11]

The mixtures of smoke and fog that settled in over London had been common in Victorian times. In 1905 Des Voeux proposed that this mixture be termed 'smog', a word with such wonderful assonance that it has never been dropped; rather it has been applied with ever-increasing frequency even to phenomena that are neither smoke nor fog (for instance the photochemical 'smog' of Los Angeles). Through the first half of the century it appeared that the famous London smogs had become rarer and the increased death rates they brought were remembered only in medical histories. It was true that there were one or two bad periods of fog, but they were mere hints of the fogs of old.

It seemed that Robert Barr's prediction of 'The Doom of London' would forever remain mere fiction, but that proved not entirely so, for London was to have its Great Smog. Reminiscent of Barr's prophesy, the weather in the week that preceded the smog of 1952 was relatively good. Each day had its gentle breezes and glimpses of sunshine, but by Thursday 4 December the conditions began to deteriorate. The winds became slacker, the air damper and the skies grey. A slow-moving anticyclone came to a halt over the city of London. By Thursday evening it was evident that London would be very foggy.

When Friday came the scene was positively Dickensian. There was

> fog everywhere, fog up the river where it flows among green aits and meadows – fog down the river, where it rolls defiled among the tiers of shipping and the waterside pollutions of a great (and dirty) city. Fog on the Essex marshes, fog on the Kentish heights. Fog creeping into the cabooses of collier-brigs; fog lying out on the yards, and hovering in the rigging of great ships; fog drooping on the gun-whales of barges and small boats. Fog in the eyes and throats of ancient Greenwich pensioners, wheezing by the firesides of their wards; fog in the stem and bowl of the afternoon pipe of the wrathful skipper, down in his close cabin; fog cruelly pinching the toes and fingers of his shivering little 'prentice boy on the deck.[12]

The fog was thicker on that Friday morning than many people could ever remember. Through the day it steadily grew even thicker. In the afternoon people were already experiencing discomfort, and noticing the choking smell in the air. Those who walked about in the fog found their skin and clothing quite filthy after a short time. By Friday night the treatment of respiratory cases was running at twice its normal level and the anticyclone had stalled completely. A million chimneys poured smoke out into the foggy stagnant air. It became ever more polluted as Londoners tried to dispel the cold and gloom.

On Saturday the fog was still there. No breezes had come to drive it off. Gradually, with visibility near zero, the transport system began to grind to a halt. People continued to suffer and some died. As with the fog of 1873, prize show animals had to be destroyed. On Sunday the fog continued and so did the deaths. The emergency services were no longer able to respond in any effective way. It is doubtful whether many people perceived the nature of the calamity that had befallen them. The Victorians would have known that such fogs were killers, but they had become uncommon in the twentieth century. When Monday morning came conditions seemed

FIGURE 8.2 Photograph of the London smog of 1952

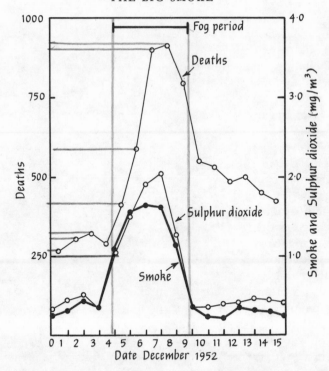

FIGURE 8.3 Deaths and pollutant concentrations during the London smog of 1952

slightly better and the transport services gradually came to life, although delays abounded. On Tuesday the Great Smog was over.

It was difficult to describe quite what had happened. The air pollution monitoring equipment operating in London was still fairly primitive and primarily designed to give long-term measurements of the air pollutant concentrations. It was not particularly suited to show the rapid changes that had occurred throughout the incident. This meant that some people believed that the official measurements were inaccurate and others believed that they had been tampered with. Despite this the measurements published since the smog seem to give a coherent picture of the incident. The changes in sulphur dioxide and smog concentrations are compared with mortality in Fig. 8.3. The values here represent the average for twelve different London stations, so some locations would indeed have experienced higher levels of smoke. The highest daily mean recorded was 4460 μg/m^3, but over shorter periods it would have been much higher. The filters of the air-conditioning system of the National Gallery normally clog at a fairly slow rate with the accumulation of particulate material from the London air. In one day during the Great Smog, the filters clogged at 26 times the normal rate and in one four-hour period they clogged at 54 times the normal rate. If one assumes that the normal loading of the London air was about 250 $\mu g/m^3$, then the four hours of densest smoke at the National Gallery could have been as high as 14,000 $\mu g/m^3$: a phenomenally high level of pollution, though it is possible that the flocculating properties of the particles were affected by the fog. The highest hourly value recorded in later air pollution incidents in London was 7200 $\mu g/m^3$.

When Parliament gathered after the Christmas recess, government ministers were subjected to a barrage of questions. There seemed little enthusiasm on the part of the government for new legislation and ministers drew attention to the powers local bodies had under the Public Health Act 1936. The smog could not be ignored, however. It became the subject of investigations by the Beaver Committee in 1953, whose final report appeared in 1954. It was not particularly innovative, but important nevertheless, as it drew together much of the thinking on smoke abatement that had accumulated over the previous decades. Continual pressure in the months that followed meant that the government was not allowed to neglect the report.

Possible government lethargy in enacting the recommendations of the Beaver Committee was forestalled when a Clean Air bill was privately introduced into the House of Commons by Gerald Nabarro, only to be withdrawn when it was clear that the government was committed to a bill of its own. The government's bill was debated in late 1955. It was criticized both by the Opposition and by Nabarro as being too weak. The hand of the Federation of British Industry was claimed to be at work in its drafting. It gave industry seven years before complete compliance was necessary. 'Practicability' and 'reasonableness' continued to be dominating features of the legislation. Local authorities would not be compelled to create smokeless zones. During the elections, both parties supported air pollution reform and the new government enacted the Clean Air Act of 5 July 1956.

Cleaner air

Perhaps the most radical element of the Clean Air Act of 1956 was that for the first time legislation attempted to control domestic sources of pollution as well as those of industry. The law was still restricted to smoke, but prohibited dark smoke. The problem of black smoke, dark smoke and smoke of other shades has arisen before, so this time it was to have a definition. Dark smoke was to be defined as anything darker than lattice two on the Ringelmann chart (Fig. 8.4). As can be seen in the long struggle for cleaner air, there was a need to raise the public awareness of its importance. It had been no use for a few idealistic activists to press for change: it had to be supported by the public at large. After all, public liberty is infringed by the desire for cleaner air, so it is necessary that the

FIGURE 8.4 Ringelmann charts. The shade of smoke may be compared
with that of the chart by having them held at some distance so that they
take on an even grey tone. White is given the Ringelmann number 1 and
black 5

population at large is aware that the sacrifices are small compared to the
great advantages of a cleaner atmosphere. The smog of 1952 had made the
disadvantages of polluted air so obvious that it was clear to the political
parties as they fought the election that action on the issue would receive
broad public support.

Naturally it was not simply idealism that finally won over a sceptical
public and Parliament. There had been a great deal of enthusiasm for
change among the smoke abatement societies of the nineteenth century.
Even the smog of 1952 would not have been a successful ally, had it not
been for the changing social conditions. If servants had still been available
to clean out dirty grates and stoves, who would have pressed for change?
Had electricity or gas not been reasonably priced, who would have pressed
for colder homes by banning the fuel which would heat them cheaply and
efficiently?

The Clean Air Acts enabled local government to set up smoke control areas (often called smokeless zones), and it is within these areas that the emission of dark smoke from domestic and industrial sources may be restricted. In London more than 90 per cent of the city is covered by smoke control. There has been remarkably little resistance to the implementation of the Act. In England 294 areas were termed 'black' by the Beaver Committee, and in urgent need of smoke control. By the time of local government reorganization in 1974 all but fourteen local authorities had taken some steps towards implementing the Act. The creation of smokeless zones was the responsibility of local government and was not dictated from Whitehall, as local co-operation was essential. In general there was enthusiasm for the new legislation, but there were delays. Shortages of smokeless fuel, civic lethargy and the impracticality of denying miners their free stint of coal are all cited as hampering the initiation of smokeless zones at a local level. Anyone affected by proposed smoke control legislation may appeal to the Secretary of State, but these appeals are rare and are even more rarely upheld.[13] Today the total amount of smoke in the air has been reduced by 80 per cent of the level at the time of the Great Smog. Reduction in smoke has been brought about through industrial controls, where smoke-arresting equipment of a high standard and tall stacks have been very effective. Domestic sources now account for more than 90 per cent of the smoke in the air, but even so, these domestic emissions have declined considerably in the last twenty years.

While the domestic user must switch to a smokeless fuel or use electricity, the Act requires that new industrial furnaces must 'so far as practicable' be smokeless. The definition of practicability is left to the local authority and if necessary the courts; but this has not caused much difficulty in practice. There are a few exemptions under the Act. Smoke is permitted for a brief period when a furnace is being started from cold or where refuse is being burnt on demolition sites.

It is quite evident that this Act was restricted to smoke, despite the knowledge that much of the damage caused by pollution in the British Isles arose from sulphur dioxide. Thus it would seem that no steps were taken to legislate against the emission of sulphur dioxide. There are no simple methods of removing sulphur dioxide from stack gases, so it is not practical to expect domestic chimneys to be equipped to remove sulphur from their emissions. Even in the case of large industrial emissions, removal of sulphur dioxide has, up to now, not been frequent. However the picture was not totally gloomy, because the introduction of smoke control zones did lower the domestic emissions of sulphur dioxide, particularly where there was a switch to electricity, gas or low-sulphur oils. Recently the solid fuels to be used within smokeless zones have been chosen more carefully and attention has been given to their sulphur content, despite the fact that this consideration is not a statutory requirement – an example, albeit rare

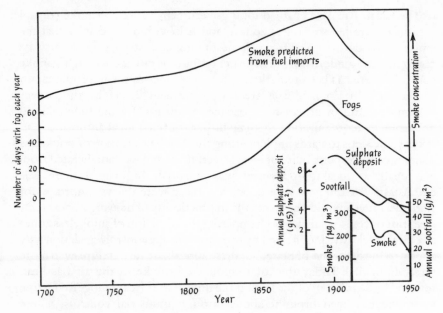

FIGURE 8.5 Air pollution in London since the seventeenth century,
comparing predicted values with fogs and later measurements

some would argue, where legislative flexibility operated to the advantage
of the environment.

The improvement in London's air has been dramatic, to say the very
least. Much of the available information on the long-term changes in
pollutant levels is summarized in Fig. 8.5. These include estimated values
for the air pollution as modelled from economic data back into the
eighteenth century, measurements of fog, and, from more recent times,
actual measurements of pollutants. The picture is an encouraging one,
suggesting that given sufficient will society can be both more affluent and
less polluted. London, once one of the 'black' areas on the Beaver map, the
foggiest city of literature and the dirtiest metropolis of Europe, may no
longer deserve such extreme appellations. Although London still has the
occasional bad fog, many small towns, considered quite clean, are now
sometimes smokier than the capital.[14]

However, there is no reason for complacency. London's enormous size
and energy-use mean that it still experiences sulphur levels higher than
many European cities. Despite increasing movement towards low-sulphur
fuels and natural gas, the levels of sulphur dioxide in the air are unlikely to
decline much further, in fact they could grow worse if there is any shift
away from the use of low-sulphur fuels; and such a shift could occur with
the increasing energy shortages. The magnitude of London's emissions is
emphasized by the results of a recently completed survey by the Scientific

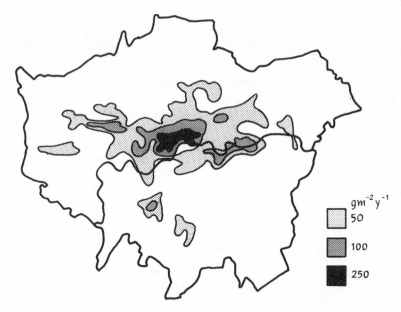

FIGURE 8.6 The distribution of sulphur emissions in London during the early 1980s

Branch of the Greater London Council.[15] In the year beginning April 1975 no less than 179,000 tonnes of sulphur dioxide were emitted from combustion processes within Greater London. The distribution of emissions may be seen in Fig. 8.6.

The fact that the control of sulphur dioxide was not an explicit part of the Clean Air Act of 1956 meant that there was a need for a body of legislation to cover it and, for that matter, atmospheric pollutants in general. In 1972 the Corporation of the City of London obtained under the City of London (Various Powers) Act the right to limit the sulphur content of fuel within the 3 square kilometres of the City of London to 1 per cent. A national law to cover sulphur dioxide in the atmosphere never arose, because it was pre-empted by the Control of Pollution Act 1974, which recognized the environment as a single entity, devoting separate control considerations to the various areas of air pollution, water pollution, waste disposal and noise nuisance.[16] Although this Act represents an improvement over the Clean Air Act 1956, it is being 'adopted into' law only slowly. The recommendations of the Royal Commission on Environmental Pollution go even further than the Control of Pollution Act. They retain the notion of 'best practicable means' that typifies British policy. Although the Commission was against setting air-quality standards as a specific framework for air pollution control, the policy advocated the need for air-quality guidelines

to aid decision-making. Perhaps even more radical was the recommend-
ation that the Alkali Inspectorate be subsumed into a new central body, Her
Majesty's Pollution Inspectorate, which would be able to cope with the
increasing complexity of the environmental problem and to search for the
best practicable environmental option.

The rate at which ideas on improving the quality of urban air became
incorporated into environmental law is slow at best. Fortunately some local
governments do not wait to be provoked by legislation. The Greater
London Council set up air-quality guidelines of its own and these are
currently used in research and planning. As we saw above, the City of
London has already limited the amount of sulphur in fuels to less than 1 per
cent.[17] A number of cities such as Coventry and Norwich have set up their
own Pollution Prevention Panels: these are voluntary bodies which
encourage informal meetings between public health officials, industrial
chemists and academics to exchange views on local pollution problems.

In the future, environmental legislation may well develop within the
European Parliament,[18] which has become increasingly concerned about
environmental pollution. In particular, as air masses have no respect for
political boundaries, air pollution has been a matter of international
concern. So far there has been more conflict than progress within the
European Commission, with the UK staunchly defending its own practical
philosophy against what is seen as continental idealism. It seems inevitable
that central governments will increasingly set standards for levels of
pollutants allowed in the air and in effluent, although to what extent these
become guidelines or remain as enforceable legal limits is not clear.

The Commission has adopted draft directives on the sulphur content of
fuels, on air-quality standards for lead, smoke and sulphur dioxide, on
emissions from motor vehicle engines and on the lead content of petrol.
This legislation demands tighter controls over pollution than the laws
already in force. The sulphur dioxide levels are becoming increasingly
difficult and expensive to lower further and they will probably need to
receive the most attention if the European directives are to be met. Laws
that originate within the UK might not be quite as stringent on sulphur
dioxide emissions as Continental legislation. However, there will continue
to be pressure from the Continent for the UK to lower them.

Smoke and more recently sulphur dioxide seem to have been the only
major concern of those who have advocated cleaner air. This is not without
some justification, as these pollutants have certainly given rise to some of
the most serious problems in the air over the cities of England. Some of the
more uncommon compounds that are released into the air today are really a
function of our increasingly sophisticated technology, yet there is a notable
problem that seems to have received little attention despite the fact that it
has been with us since the earliest times: odour control. This can be
extremely difficult and there are cases where, even after the application of

the best practicable means, plants have had to be closed because they were unable to reduce odour to acceptable levels. Odour is notoriously difficult to monitor and defining harmful odours is obviously even more troublesome. The smell of chocolate or beer wafting over across the suburbs may not sound very unpleasant, but some otherwise unpolluted locations are marred by unacceptable levels of olfactory assault.[19]

Other air pollutants

Carbon monoxide arises mostly from the incomplete combustion of fuels in automobile engines. It is odourless, colourless and poisonous. The toxicity arises from the fact that the carbon monoxide binds very strongly with the haemoglobin in the blood, preventing it from carrying oxygen to the tissues. In rush-hour traffic 20–30 ppm concentrations are to be expected in London air. Such concentrations, if prolonged, might lead to some temporary impairment in ability to discriminate sound, sight or time periods. Headaches are also a result of exposure to this toxic gas. However, the highest blood levels of carbon monoxide are usually the result of smoking rather than automotive pollution.[20]

Lead is found in concentrations as high as 4.1 $\mu g/m^3$ in the London air, but the exact physiological effect of such levels are not known. Lead in the environment seems to be ingested by children, and it is here that there is the greatest potential for damage. The lead in the urban environment originates from the combustion of motor fuels which contain lead tetra-ethyl as an anti-knock agent. However, in line with an EEC directive, the Motor Fuel (Lead Content of Petrol) Regulations 1976 required that from 1977 petrol should contain no more than 0.45 g/l. The EEC proposals also require that the mean annual levels in urban residential areas should not exceed 2 $\mu g/m^3$. In general it seems that there will be no difficulty in UK cities meeting this standard if it is adopted.[21]

Perhaps even more worrying than any of the pollutants mentioned above are a new set of contaminants that are actually being generated in the air over London. Although pollutants produced in photochemical processes in the atmosphere have probably always been in London's air in small quantities, the very process of cleaning up the air so desired by the Clean Air Act has contributed to increased pollution of this kind. The cleaner air has meant more sunlight, and hence more photo-oxidation of the contaminants of the urban air. London has begun to experience the smogs usually associated with Los Angeles.[22]

At the root of the photochemical smog problem are the nitrogen oxides NO and NO_2 which are sometimes written in sum as NO_x. Nitrous oxide (NO) arises from the oxidation of atmospheric nitrogen drawn in with the air during combustion. The automobile represents a very important urban source of this gas. Nitrous oxide is liable to further oxidation, that may be rapid in the presence of hydrocarbons (from unburnt fuel) and light:

$$CH_4 + 2O_2 + 2NO + light \rightarrow H_2O + HCHO + 2NO_2$$

In the equation above methane (CH_4) has been used as an illustrative example of an atmospheric hydrocarbon. It has been oxidized to formaldehyde (HCHO) in the photochemical reaction. Formaldehyde is a suspect carcinogen, so hardly a desirable atmospheric component. The oxidation of nitrous oxide (NO) through to nitrogen dioxide (NO_2) shifts the delicate equilibrium that exists between these two nitrogen oxides and ozone in the atmosphere. This forces the ozone concentrations to much higher levels.

The ozone measurements made in the London atmosphere over recent years cover too short a time period to draw definite conclusions, but there is cause for concern. The summer of 1976 was not only notable for its lack of rain. London experienced the worst episodes of photochemical smogs in its history. On 27 June an hourly mean for ozone of 21.2 pphm (parts per hundred million) was recorded. This exceeds the 8 pphm guideline used by the Greater London Council by a considerable margin. During the summer of 1976 the guideline was exceeded at one or more of the sites in Greater London on 37 per cent of the days. The high ozone concentrations are particularly worrying in the light of some suggestions that the high levels of ozone now occasionally found in the London air may exacerbate the effect of sulphur dioxide on the lungs of sensitive individuals.[23]

Some of the nitrogen oxides in the urban air may react with partially oxidized organic compounds in the atmosphere to form compounds such as PAN which is the major irritant present in photochemical smog (PAN is properly known as peroxyacetylnitrate, $CH_3COO_2NO_2$). It may also be a carcinogenic. In the summer of 1976 reports of eye irritation from PAN and related photochemical pollutants were well correlated with the occurrence of high levels of ozone in the atmosphere. Photochemical oxidants cause damage to vegetation, although the extent of this damage during the 1976 episodes is not known. However, the spread of photochemical pollutants out from London may be significant as far as 100 km downwind. The high levels of ozone also cause damage to materials. Rubber is cracked, and although there is no conclusive evidence for this in London yet, there have been indications that some synthetic fabrics are discoloured, reminding us of the suggestions of Robert Boyle some three centuries earlier.

The highly oxidizing components of the photochemical smog can oxidize the sulphur dioxide present in the air to sulphuric acid. This, along with nitric acid, an oxidation product of the nitrogen oxides, can produce a haze which lowers visibility. Such an atmospheric haze was frequently observed during the photochemical episodes in 1976. In addition to the haze, on some days a brownish hue could be observed close to the horizon.

The precursors to the photochemical pollutants that are produced in the atmosphere come predominantly from cars. The private vehicle is obviously also implicated in the spread of lead through the urban atmosphere. So control of this particular source of pollution will represent yet another infringement on the apparent freedom of the individual.

In the future financial considerations and increasing pressure on limited energy resources can be expected to make it ever harder to keep the urban atmosphere unpolluted. But no one can imagine a return to the days of 'London's Particular', however nostalgic we feel. Continued public awareness will ensure such things are prevented. As the range of materials needed by modern technology broadens, so do the dangers of air pollution by materials whose effects within the complex matrix of the urban atmosphere are unknown. It is tragic, but true, that dangers are often impossible to predict. Where anticipation has been absent, a willingness to respond to new problems as quickly as possible is obviously important. Let us hope that improvements continue without the need for something like a *Great Photochemical Smog* to stir the imagination.

Notes

1. Marsh, A. (1947) *Smoke*, Faber & Faber, London.
2. Fitzgerald, (1939–40) 'Report on investigation on the cost of smoke', *Smokeless Air*, 10, (39–40). The cost of cleaning silverware was noted by Russell, R. (1888) *Smoke in Relation to Fogs in London*, National Smoke Abatement Institute, and by Digby, Sir K. (1658) *A Discourse on Sympathetic Powder* much earlier.
3. London Underground engineers will readily explain about the blackening of the tunnels by the steam trains to the willing listener. Pollution from steam trains is a subject in itself; a nice little discussion can be found in Alcock, G. R. (1949) 'Is the brick arch necessary?', *The Model Engineer*, 596–8. See also 'Smoke prevention on railways', *English Mechanic and Mirror of Science and Art*, 5 (1867), 290; Caruthers, C. H. (1905) 'Early experiments with smoke-consuming fire boxes on American locomotives', *Railroad Gazette*, 39, 514.
4. Obviously in the years before radar, smoke was a considerable hazard to aviators.
5. . . . but then clearly he stuck to the spirit of his title more than I.
6. *Evening Standard*, London 8 December 1945; *Evening Standard*, London 19 January 1946. In the same vein, but from a writer we might prefer to forget, we find Morley, C. D. (1921) *Chimney Smoke*, George H. Doran, New York:

At night I opened	The fire that sparkled
The furnace door:	Blue and red
The warm glow brightened	Kept small toes cosy
The cellar floor.	In their bed

As up the stair
So late I stole
I said my prayer:
Thank God for coal!

7. London, J. (1904) *People of the Abyss*, Macmillan, New York.
8. Anon. (1912) 'The sootfall of London: its amount, quality and effects', *The Lancet*, 47–50.
9. Shaw, N. and Owens, J. S. (1925) *The Smoke Problem of Great Cities*, Constable, London.
10. This delightful incident and much more is to be found in the third paper of Ashby, E. and Anderson, M. (1977) 'Studies in the politics of environmental pollution: the historical roots of the British Clean Air Act, 1956: III', *Interdisciplinary Science Reviews*, 2 190–206. The Englishman's desire to retain his roaring fire is also to be found in Bevan, P. (1872) 'Our national coal cellar', *Gentleman's Magazine*, NS9, 268–78, and Baines, Sir F. (1925) *J. Roy. Soc. Arts*, 73, 453.
11. The title and much of the background for this section comes from Wise, W. (1968) *Killer Smog*, Rand McNally, Skokie, IL, a semi-fictional account of experiences throughout the smog. More can be found in Ashby, E. and Anderson, M. (1981) *The Politics of Clean Air*, Oxford University Press.
12. Dickens, C. (1852–3) *Bleak House*, Bradbury & Evans, London, published in parts.
13. *Royal Commission on Environmental Pollution* (1976) 5th Report.
14. The fogs that occurred after 1952 are shown in Table 6.2, p. 114. For a worldwide comparison see de Koning, H. W., Kretzschmar, J. G., Akland, G. G. and Bennett, B. G. (1986) 'Air pollution in different cities around the world', *Atmospheric Environment*, 20, 101–13.
15 Ball, D. J. and Radcliffe, S. W. (1979) *An Inventory of Sulphur Dioxide Emissions to London's Air*, GLC Res. Report 23, GLC, London.
16. Bennett, G. (1979) 'Pollution control in England and Wales', *Environmental Policy and Law*, 5, 93–9.
17. Manifold, B. (1979) 'The European Commission and its influence on pollution control in the United Kingdom', *Environ. Health*, 87, 121.
18. ibid., and *Official Journal of the European Communities*, no. C54/79.
19. Henderson-Sellers, B. (1984) *Pollution of Our Atmosphere*, Adam Hilger Ltd, Bristol.
20. Perkins, H. C. (1975) *Air Pollution*, McGraw-Hill, New York.
21. Manifold, B. (1979) 'The European Commission and its influence on pollution control in the United Kingdom', *Environ. Health*, 87, 121.
22. Apling, A. J., Sullivan, E. J., Williams, M. L., Ball, D. J., Bernard, R. E., Derwent, R. G., Eggleton, A. E. J., Mapton, L. and Waller, R. E. (1977) 'Ozone concentrations in South-East England during the summer of 1976', *Nature*, 269, 569–73; Ball, D. J. (1978) 'Evidence of photochemical haze in the atmosphere of greater London', *Nature*, 271, 372–6.
23. Hazucha, M. and Bates, D. V. (1975) 'Combined effects of ozone and sulphur dioxide on human pulmonary function', *Nature*, 257, 50.

Annotated bibliography

There are relatively few books on air pollution history. One could do worse than to read the classic *Fumifugium*, although it is not easy to find in modern editions. On the broader subject of air pollution there are so many books that it is hard to decide which to recommend. There are numerous papers on specialized aspects of this work. They will be found referenced in the notes at the end of each chapter.

Ashby, E. and Anderson, M. (1981) *The Politics of Clean Air*, Oxford University Press. A very thorough account of the rise of air pollution legislation in England since the last century.

Brimblecombe, P. (1986) *Air Composition and Chemistry*, Cambridge University Press.

Clayre, A. (1977) *Nature and Industrialization*, Oxford University Press. Looks at the area of Romanticism, art, literature and industry.

Evelyn, J. (1661) *Fumifugium, or The Inconvenience of the Aer and Smoak of London Dissipated . . .*, printed by W. Godbid for Gabriel Bedel and Thomas Collins, London. Two modern reprints are *Fumifugium* in J. P. Lodge (ed.) (1970) *The Smoke of London: Two Prophesies*, Maxwell Reprint Co., London, and *Fumifugium*, The Rota, University of Essex, Colchester, (1976).

Galloway, R. L. (1898/1904) *Annals of Coal Mining and the Coal Trade*, reprinted David & Charles, Newton Abbot, (1971).
(1882) *History of Coal Mining in Great Britain*, reprinted David & Charles, Newton Abbot.

Henderson-Sellers, B. (1984) *The Pollution of our Atmosphere*, Adam Hilger Ltd, Bristol. An introduction to air pollution in general.

Howe, G. Melvyn (1972) *Man, Environment and Disease in Britain*, David & Charles, Newton Abbot. A medical geography through the ages.

Nef, J. U. (1932) *The Rise of the British Coal Trade*, Routledge & Kegan Paul, London, remains available in reprints and is still a magnificent work.

Wise, W. (1968) *Killer Smog*, Rand McNally, skogie, IL. A semi-fictional account of the London smog of 1952.

Index